T0192793

Data-Centric Systems and Applications

Series editors

Michael J. Carey
Stefano Ceri

Editorial Board

Anastasia Ailamaki
Shivnat Babu
Philip Bernstein
Johann-Christoph Freytag
Alon Halevy
Jiawei Han
Donald Kossmann
Ioana Manolescu
Gerhard Weikum
Kyu-Young Whang
Jeffrey Xu Yu

More information about this series at http://www.springer.com/series/5258

George Fletcher • Jan Hidders •
Josep Lluís Larriba-Pey

Editors

Graph Data Management

Fundamental Issues and Recent
Developments

 Springer

Editors
George Fletcher
Department of Mathematics
and Computer Science
Eindhoven University of Technology
Eindhoven, The Netherlands

Jan Hidders
Department of Computer Science
Vrije Universiteit Brussel
Brussels, Belgium

Josep Lluís Larriba-Pey
Department of Computer Architecture
Universitat Politècnica de Catalunya
Barcelona, Spain

ISSN 2197-9723   ISSN 2197-974X   (electronic)
Data-Centric Systems and Applications
ISBN 978-3-030-07151-6   ISBN 978-3-319-96193-4   (eBook)
https://doi.org/10.1007/978-3-319-96193-4

© Springer International Publishing AG, part of Springer Nature 2018
Softcover reprint of the hardcover 1st edition 2018
This work is subject to copyright. All rights are reserved by the Publisher, whether the whole or part of the material is concerned, specifically the rights of translation, reprinting, reuse of illustrations, recitation, broadcasting, reproduction on microfilms or in any other physical way, and transmission or information storage and retrieval, electronic adaptation, computer software, or by similar or dissimilar methodology now known or hereafter developed.
The use of general descriptive names, registered names, trademarks, service marks, etc. in this publication does not imply, even in the absence of a specific statement, that such names are exempt from the relevant protective laws and regulations and therefore free for general use.
The publisher, the authors and the editors are safe to assume that the advice and information in this book are believed to be true and accurate at the date of publication. Neither the publisher nor the authors or the editors give a warranty, express or implied, with respect to the material contained herein or for any errors or omissions that may have been made. The publisher remains neutral with regard to jurisdictional claims in published maps and institutional affiliations.

This Springer imprint is published by the registered company Springer Nature Switzerland AG
The registered company address is: Gewerbestrasse 11, 6330 Cham, Switzerland

Preface

The area of graph data management has recently seen many exciting and impressive developments. It addresses one of the great scientific and industrial trends of today: leveraging complex and dynamic relationships to generate insight and competitive advantage. It is crucial for such different goals as understanding relationships between users of social media, customers, elements in a telephone or data center network, entertainment producers and consumers, or genes and proteins. As part of the NoSQL movement it provides us with new powerful technologies and means for storing, processing, and analyzing data. It also is a key technology for supporting the Semantic Web and Linked Open Data.

As a consequence, there has been an impressive flurry of new systems for graph storage and graph processing, both in academia and industry, and covering a wide spectrum of use cases, from enterprise-scale datasets to web-scale datasets. Moreover, this has been accompanied by exciting new research, developing further the foundations of efficient graph processing, as well as exploring new application areas where these can be successfully applied.

The present volume collects and presents an overview of recent advances on fundamental issues in graph data management to allow researchers and engineers to benefit from these in their research. The chapters are contributed by leading experts in the relevant areas, presenting a coherent overview of the state of the art in the field.

The aim of this book is to give beginning researchers in the area of graph data management, or in an area that requires graph data management, an overview of the latest developments in this area, both in applied and in fundamental subdomains. The main emphasis of the book is on presenting comprehensive overviews, rather than in-depth treatment of subjects, although technological subjects are not avoided. Our hope is that beginning researchers will be better positioned to take more informed decisions about their research direction or, if it is already under way, to better put their work into context. For researchers not in the domain itself, but interested in using the results from this domain, we hope this volume will help them to better understand what types of tools, techniques, and technologies are available and which ones would best suit their needs. The prerequisites for the book are a basic understanding of data management techniques as they are taught in academic computer science MSc programs.

The contributions to this volume have their genesis as lecture notes distributed to students at the 12th EDBT Summer School on "Graph Data Management," held the week of August 31, 2015 in Palamos, Spain. The school was organized by the editors of this book, under the auspices of the EDBT Association, a leading international nonprofit organization for the promotion and support of research and progress in the fields of databases and information systems technology and applications. These notes were already single-blind reviewed by the scientific committee of the summer school, before distribution to the students. All contributions to the present volume were further extended based upon experiences at the school and again subject to further editorial improvement and single-blind peer review by members of the scientific committee.

We thank the following colleagues for their service on the scientific committee of the summer school and as reviewers of the contributions to this volume.

- Paolo Atzeni. *Università Roma Tre.*
- Alex Averbuch. *Neo Technology.*
- Sourav Bhowmick. *Nanyang Technological University.*
- Angela Bonifati. *Université Claude Bernard Lyon 1 and CNRS.*
- Andrea Calì. *Birkbeck, University of London.*
- Mihai Capotă. *Intel Labs.*
- Ciro Cattuto. *ISI Foundation.*
- David Gross-Amblard. *Université de Rennes 1.*
- Olaf Hartig. *Linköpings Universitet.*
- Meichun Hsu. *Hewlett Packard Labs.*
- H.V. Jagadish. *University of Michigan.*
- Aurélien Lemay. *Université Lille 3 and INRIA.*
- Ulf Leser. *Humboldt-Universität zu Berlin.*
- Federica Mandreoli. *Università degli Studi di Modena e Reggio Emilia.*
- Thomas Neumann. *Technische Universität München.*
- Paolo Papotti. *EURECOM Sophia Antipolis.*
- Arnau Prat. *Sparsity Technologies.*
- Pierre Senellart. *École normale supérieure.*
- Sławek Staworko. *Université Lille 3 and INRIA.*
- Letizia Tanca. *Politecnico di Milano.*
- Alex Thomo. *University of Victoria.*
- Maurice Van Keulen. *Universiteit Twente.*
- Stijn Vansummeren. *Université libre de Bruxelles.*
- Ana Lucia Varbanescu. *Universiteit van Amsterdam.*
- Jim Webber. *Neo Technology.*
- Peter Wood. *Birkbeck, University of London.*

- Yuqing Wu. *Pomona College.*
- Yinglong Xia. *Huawei Research America.*

Eindhoven, The Netherlands   George Fletcher
Brussels, Belgium   Jan Hidders
Barcelona, Spain   Josep Lluís Larriba Pey

Contents

Contributors

Renzo Angles Department of Computer Science, Universidad de Talca, Curicó, Chile – and – Millennium Institute for Foundational Research on Data, Santiago, Chile

Angela Bonifati Université Claude Bernard Lyon 1, Villeurbanne, France

Peter Eades University of Sydney, Sydney, NSW, Australia

Alfredo Ferro University of Catania, Catania, Italy

George Fletcher Eindhoven University of Technology, Eindhoven, the Netherlands

Rosalba Giugno University of Catania, Catania, Italy

Claudio Gutierrez Universidad de Chile, Santiago, Chile – and – Millennium Institute for Foundational Research on Data, Santiago, Chile

Jan Hidders Vrije Universiteit Brussel, Brussels, Belgium

Alexandru Iosup Vrije Universiteit Amsterdam, Amsterdam, the Netherlands Delft University of Technology, Delft, the Netherlands

Karsten Klein Monash University, Melbourne, VIC, Australia – and – University of Konstanz, Konstanz, Germany

Yatao Li Microsoft Research Asia, Beijing, China

Giovanni Micale University of Catania, Catania, Italy

Misael Mongioví University of Catania, Catania, Italy

Alexandra Poulovassilis Birkbeck, University of London, London, UK

Alfredo Pulvirenti University of Catania, Catania, Italy

Bin Shao Microsoft Research Asia, Beijing, China

Dennis Shasha Courant Institute of Mathematical Science, New York University, New York, NY, USA

Chapter 1
An Introduction to Graph Data Management

Renzo Angles and Claudio Gutierrez

Abstract Graph data management concerns the research and development of powerful technologies for storing, processing and analyzing large volumes of graph data. This chapter presents an overview about the foundations and systems for graph data management. Specifically, we present a historical overview of the area, studied graph database models, characterized essential graph-oriented queries, reviewed graph query languages, and explore the features of current graph data management systems (i.e. graph databases and graph-processing frameworks).

1.1 Introduction

Graphs are omnipresent in our lives and have been increasingly used in a variety of application domains. For instance, the web contains tens of billions of web pages over which the page rank algorithm is computed, Facebook has billions of users whose billions of relationships are explored by social media analysis tools, and Twitter contains hundreds of millions of users whose similar amount of tweets per day are analyzed to determine trending topics. The data generated by these applications can be easily represented as graphs characterized for being large, highly interconnected and unstructured. To meet the challenge of storing and processing such big graph data, a number of software systems have been developed. In this chapter, we concentrate on graph data management systems.

Graph Data Management concerns the research and development of powerful technologies for storing, processing and analyzing large volumes of graph data (Sakr

R. Angles (✉)
Department of Computer Science, Universidad de Talca, Curicó, Chile – and – Millennium Institute for Foundational Research on Data, Santiago, Chile
e-mail: rangles@utalca.cl

C. Gutierrez
Department of Computer Science, Universidad de Chile, Santiago, Chile

Millennium Institute for Foundational Research on Data, Santiago, Chile
e-mail: cgutierr@dcc.uchile.cl

© Springer International Publishing AG, part of Springer Nature 2018
G. Fletcher et al. (eds.), *Graph Data Management*, Data-Centric Systems and Applications, https://doi.org/10.1007/978-3-319-96193-4_1

and Pardede 2011). The research on graph databases has a long development, at least since the 1980s. But it is only recently that several technological developments have made it possible to have practical graph database systems. Powerful hardware to store and process graphs, powerful sensors to record directly the information, powerful machines that allow to analyze and visualize graphs, among other factors, have given rise to the current flourishing in the area of graph data management.

We devise two broad and interrelated topics in the area of graph data management that in our opinion deserve to be treated separately today. One is the area of graph database models, which comprises general principles that ideally should guide the design of systems. The second is graph data management systems themselves, which are systems that deal with graph data storing and querying, sometimes addressing directly demands of users, thus emphasizing factors such as efficiency, usability and direct solutions to urgent data management problems.

1.1.1   Graph Database Models

The fundamental abstraction behind a database system is its database model. In the most general sense, a *database model* is a conceptual tool used to model representations of real-world entities and the relationships among them. As is well known, a database model can be characterized by three basic components, namely, data structures, query and transformation language, and integrity constraints. In the context of graph data management, a *graph database model* is a model where data structures for the schema and/or instances are modeled as graphs, where the data manipulation is expressed by graph-oriented operations, and appropriate integrity constraints can be defined over the graph structure.

1.1.2   Graph Data Management Systems

There are two categories of graph data management systems: graph databases and graph-processing frameworks. The former are systems specifically designed for managing graph-like data following the basic principles of database systems, that is, persistent data storage, physical/logical data independence, data integrity and consistency. The latter are frameworks for batch processing and analysis of big graphs putting emphasis on the use of multiple machines to improve the performance These systems provide two perspectives for storing and querying graph data, each one with their own goals.

1.1.3   Contents and Organization of This Chapter

This chapter presents an overview of the basic notions, the historical evolution and the main developments in the area of graph data management. There are three main topics, distributed by sections. First, an overview of the field and its development, which we hope can be of help to look for ideas and past experiences. Second, a review of the main graph database models in order to give a perspective on actual developments. Third, a similar review of graph database query languages. Finally, we present current graph data management systems in a comparative manner.

1.2 Overview of the Field

In this section, we present motivations for graph data management and briefly review the developments thereof. There is an emphasis on models in order to give a certain abstraction level and unity of concepts that sometimes get lost in the wide diversity of syntaxes and implementation solutions that exist today. This section follows closely the survey of graph database models written by Angles and Gutierrez (2008).

1.2.1 What is a Graph Database Model?

A *Graph Database Model* is a model in which the data structures for the schema and/or instances are modeled as a directed, possibly labeled, graph or generalizations of the graph data structure, where data manipulation is expressed by graph-oriented operations and type constructors, and appropriate integrity constraints can be defined over the graph structure (Angles and Gutierrez 2008).

The main characteristic of a graph database model is that the data are conceptually modeled and presented to the user as a graph, that is, the data structures (data and/or schema) are represented by graphs, or by data structures generalizing the notion of graph (e.g., hypergraphs or hypernodes). One of the main features of a graph structure is the simplicity to model unstructured data. Therefore, in graph models the separation between schema and data (instances) is less marked than in the classical relational model.

Regarding data manipulation and querying, it is expressed by graph transformations, or by operations whose main primitives are based on graph features like paths, neighborhoods, subgraphs, graph patterns, connectivity, and graph statistics (diameter, centrality, etc.). Some graph models define a flexible collection of type constructors and operations, which are used to create and access the graph data structures. Another approach is to express all queries using a few powerful graph manipulation primitives. Usually the query language is what gives a database model its particular flavor. In fact, the differences among graph data structures are usually minor as compared to differences among graph query languages.

Finally, integrity constraints enforce data consistency. These constraints can be grouped in schema–instance consistency, identity and referential integrity, and functional and inclusion dependencies. Examples of these are labels with unique names, typing constraints on nodes, functional dependencies, domain and range of properties, and so on.

1.2.2 Historical Overview

The ideas of graph databases can be dated at least to the 1990s, where much of the theory developed. Probably due to the lack of hardware support to manage big graphs, this line of research declined for a while until a few years ago, when processing graphs became common and a second wave of research was initiated.

1.2.2.1 The First Wave

In an early approach, facing the failure of contemporary systems to take into account the semantics of a database, a semantic network to store data about the database was proposed by Roussopoulos and Mylopoulos (1975). An implicit structure of graphs for the data itself was presented in the Functional Data Model (Shipman 1981), whose goal was to provide a "conceptually natural" database interface. A different approach proposed the Logical Data Model (Kuper and Vardi 1984), where an explicit graph data model intended to generalize the relational, hierarchical and network models. Later, Kunii (1987) proposed a graph data model for representing complex structures of knowledge called G-Base.

GOOD (Gyssens et al. 1990) was an influential graph-oriented object model, intended to be a theoretical basis for a system in which manipulation as well as representation are transparently graph-based. Among the subsequent developments based on GOOD are: GMOD (Andries et al. 1992), which proposes a number of concepts for graph-oriented database user interfaces; Gram (Amann and Scholl 1992), which is an explicit graph data model for hypertext data; PaMaL (Gemis and Paredaens 1993), which extends GOOD with explicit representation of tuples and sets; GOAL (Hidders and Paredaens 1993), which introduces the notion of association nodes; G-Log (Paredaens et al. 1995), which proposed a declarative query language for graphs; and GDM (Hidders 2002), which incorporates representation of n-ary relationships.

There were proposals that used generalization of graphs with data modeling purposes. The Hypernode Model (Levene and Poulovassilis 1990) was a model based on nested graphs on which subsequent work was developed (Poulovassilis and Levene 1994; Levene and Loizou 1995). The same idea was used for modeling multiscaled networks (Mainguenaud 1992) and genome data (Graves et al. 1995). Another generalization of graphs, hypergraphs, gave rise to another family of models. GROOVY (Levene and Poulovassilis 1991) is an object-oriented data model based on hypergraphs. This generalization was used in other contexts: query and visualization in the Hy+ system (Consens and Mendelzon 1993); modeling of data instances and access to them (Watters and Shepherd 1990); representation of user state and browsing (Tompa 1989).

There are several other proposals that deal with graph data models. Güting (1994) proposed GraphDB, intended for modeling and querying graphs in object-oriented databases and motivated by managing information in transport networks. Database Graph Views (Gutiérrez et al. 1994) proposed an abstraction mechanism to define and manipulate graphs stored in either relational, object-oriented or file systems. The project GRAS (Kiesel et al. 1996) uses attributed graphs for modeling complex information from software engineering projects. The well-known OEM (Papakonstantinou et al. 1995) model aims at providing integrated access to heterogeneous information sources, focusing on information exchange.

Another important line of development has to do with data representation models and the World Wide Web. Among them are data exchange models like XML (Bray et al. 1998), metadata representation models like RDF (Klyne and Carroll 2004) and ontology representation models like OWL (McGuinness and van Harmelen 2004).

1.2.2.2 The Second Wave

We are witnessing the second impulse of development of graph data management, which is focused, on one hand, in practical systems and on the other, in theoretical analyses particularly of graph query languages. We will review the former in Sect. 1.5 concentrating on database systems, and will leave the latter out of this chapter. With regard to modern graph query languages the interested reader can read the tutorial of Barceló Baeza (2013) and the survey of Angles et al. (2017).

1.2.3 Comparison with Classical Models

As is well known, there are manifold approaches to model information and knowledge, depending on application areas and user needs. The first question one should answer is why choose a graph data model instead of a relational, object-oriented, semi-structured, or other type of data model. The one-sentence answer is: Graph models are designed to manage data in areas where the main concern has to do with the interconnectivity or topology of that data. In these applications, the atomic data and the relations among the units of data have the same level of importance.

Among the main advantages that graph data models offer over other types of models, we can mention the following:

- Graphs have been long recognized as one of the most simple, natural and intuitive knowledge representation systems. This simplicity overcomes the limitations of the linear format of classical writing systems.

- Graph data structures allow for a natural modeling when data have a graph structure. Graphs have the advantage of being able to keep all the information about an entity in a single node and show related information by arcs connected to it. Graph objects (like paths, neighborhoods) may have first-order citizenship.
- Queries can address directly and explicitly this graph structure. Associated with graphs are specific graph operations in the query language algebra, such as finding shortest paths, determining certain subgraphs, and so forth. Explicit graphs and graph operations allow users to express a query at a high level of abstraction. In summary, graph models realize for graph data the separation of concerns between modeling (the logic level) and implementation (physical level).
- Implementation-wise, graph databases may provide special graph storage structures, and take advantage of efficient graph algorithms available for implementing specific graph operations over the data.

Next, we will briefly review the most influential data models (relational, semantic, object-oriented, semistructured) and compare them to graph data models.

The *relational data model* was introduced by Codd (1970) and is based on the simple notion of relation, which together with its associated algebra and logic, made the relational model a primary model for database research. In particular, its standard query and transformation language, SQL, became a paradigmatic language for querying. It popularized the concept of abstraction levels by introducing a separation between the physical and logical levels. Gradually, the focus shifted to modeling data as seen by applications and users (i.e. tables). The differences between graph data models and the relational data model are manifold. The relational model is geared toward simple record-type data, where the data structure is known in advance (airline reservations, accounting, inventories, etc.). The schema is fixed, which makes it difficult to extend these databases. It is not easy to integrate different schemas, nor is it automatized. The table-oriented abstraction is not suitable to naturally explore the underlying graph of relationships among the data, such as paths, neighborhoods, patterns.

Semantic data models (Peckham and Maryanski 1988) focus on the incorporation of richer and more expressive semantics into the database, from a user's viewpoint. Database designers can represent objects and their relations in a natural and clear manner (similar to the way users view an application) by using high-level abstraction concepts such as aggregation, classification and instantiation, sub- and superclassing, attribute inheritance and hierarchies. A well-known and successful case is the entity-relationship model (Chen 1976), which has become a basis for the early stages of database design. Semantic data models are relevant to graph data model research because the semantic data models reason about the graph-like structure generated by the relationships between the modeled entities.

Object-oriented (O-O) data models (Kim 1990) are designed to address the weaknesses of the relational model in data-intensive domains involving complex data objects and complex object interactions, such as CAD/CAM software, computer graphics and information retrieval. According to the O-O programming paradigm on which these models are based, they represent data as a collection of objects that are organized into classes, and have complex values and methods. O-O data models are related to graph data models in their explicit or implicit use of graph structures in definitions. Nevertheless, there are important differences with respect to the approach for deciding how to model the world. O-O data models view the world as a set of complex objects having certain state (data), where interaction is via method passing. On the other hand, graph data models view the world as a network of relations, emphasizing data interconnection, and the properties of these relations. O-O data models focus on object dynamics, their values and methods.

Semistructured data models (Buneman 1997) were motivated by the increased existence of semistructured data (also called unstructured data), data exchange, and data browsing mainly on the web. In semistructured data, the structure is irregular, implicit and partial; the schema does not restrict the data, it only describes it, a feature that allows extensible data exchanges; the schema is large and constantly evolving; the data is self-describing, as it contains schema information. Representative semistructured models are OEM (Papakonstantinou et al. 1995) and Lorel (Abiteboul et al. 1997). Many of these ideas can be seen in current semistructured languages like XML or JSON. Generally, semistructured data are represented using a tree-like structure. However, cycles between data nodes are possible, which leads to graph-like structures as in graph data models.

1.3 Graph Database Models

All graph data models have as their formal foundation variations on the basic mathematical definition of a graph, for example, directed or undirected graphs, labeled or unlabeled edges and nodes, properties on nodes and edges, hypergraphs and hypernodes.

The most simple model is a plain labeled graph, that is, a graph with nodes and edges as everyone knows it. Although highly easy to learn, it has the drawback that it is difficult to modularize the information it represents. The notions of hypernodes and hypergraphs address this problem. Hypergraphs, by enhancing the notion of simple edge, allow the representation of multiple complex relations. On the other hand, hypernodes modularize the notion of node, by allowing nesting graphs inside nodes. As drawbacks, both models use complex data structures that make their use and implementation less intuitive.

Regarding simplicity, one of the most popularized models is the semistructured model, which uses the most simple version of a graph, namely, a tree, the most common and intuitive way of organizing our data (e.g., directories). Finally, the most common models are slightly enhanced versions of the plain graphs. One of them, the RDF model, gives a light typing to nodes, and considers edges as nodes, giving uniformity to the information objects in the model. The other, the property graph model, allows to add properties to edges and nodes.

Next, we will present these models and show a paradigmatic example of each. We will use the toy genealogy database presented in Fig. 1.1.

PERSON		
ID	NAME	LASTNAME
1	George	Jones
2	Ana	Stone
3	Julia	Jones
4	James	Deville
5	David	Deville
6	Mary	Deville

PARENT	
PERSON_ID	PARENT_ID
3	1
3	2
5	4
5	3
6	4
6	3

Fig. 1.1 A relational database of genealogical data. The table PERSON contains information about people, and the table PARENT contains pairs of people related by the children of relationship

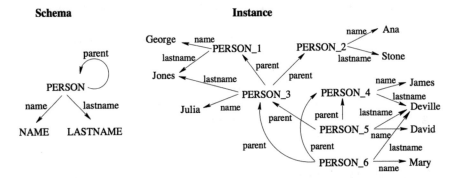

Fig. 1.2 Gram. At the schema level we use generalized names for definition of entities and relations. At the instance level, we create instance labels (e.g., PERSON_1) to represent entities, and use the edges (defined in the schema) to express relations between data and entities

1.3.1 The Basics: Labeled Graphs

The most basic data structure for graph database models is a directed graph with nodes and edges labeled by some vocabulary. A good example is Gram (Amann and Scholl 1992), a graph data model motivated by hypertext querying.

A schema in Gram is a directed labeled multigraph, where each node is labeled with a symbol called a *type*, which has associated a domain of values. In the same way, each edge has assigned a label representing a relation between types (see example in Fig. 1.2). A feature of Gram is the use of regular expressions for explicit definition of paths called *walks*. An alternating sequence of nodes and edges represents a walk, which combined with other walks forms other special objects called *hyperwalks*.

For querying the model (particularly path-like queries), an algebraic language based on regular expressions is proposed. For this purpose a hyperwalk algebra is defined, which presents unary operations (projection, selection, renaming) and binary operations (join, concatenation, set operations), all closed under the set of hyperwalks.

1.3.2   Complex Relations: The Hypergraph Model

A hypergraph is a generalization of a graph where the notion of edge is extended to *hyperedge*, which relates to an arbitrary set of nodes (Berge 1973). Hypergraphs allow the definition of complex objects by using undirected hyperedges, functional dependencies by using directed hyperedges, object-ID and multiple structural inheritance.

A good representative case is GROOVY (Levene and Poulovassilis 1991), an object-oriented data model that is formalized using hypergraphs. An example of a hypergraph schema and instance is presented in Fig. 1.3. The model defines a set of structures for an object data model: value schemas, objects over value schemas, value functional dependencies, object schemas, objects over object schemas and class schemas. The model shows that these structures can be defined in terms of hypergraphs.

Groovy also includes a hypergraph manipulation language (HML) for querying and updating hypergraphs. It has two operators for querying hypergraphs by identifier or by value, and eight operators for manipulation (insertion and deletion) of hypergraphs and hyperedges.

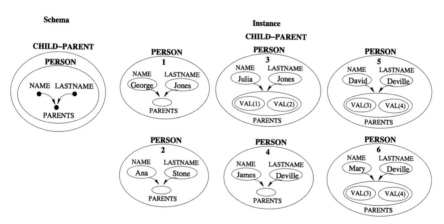

Fig. 1.3 GROOVY. At the schema level (left), we model an object *PERSON* as an hypergraph that relates the attributes *NAME, LASTNAME* and *PARENTS*. Note the value functional dependency (VFD) *NAME,LASTNAME → PARENTS* logically represented by the directed hyperedge ({NAME,LASTNAME} {PARENTS}). This VFD asserts that NAME and LASTNAME uniquely determine the set of PARENTS

1.3.3   Nested Graphs: The Hypernode Model

A hypernode is a directed graph whose nodes can themselves be graphs (or hypernodes), allowing nesting of graphs. Hypernodes can be used to represent *simple* (flat) and *complex objects* (hierarchical, composite and cyclic) as well as mappings and records. A key feature is its inherent ability to *encapsulate information*.

The hypernode model was introduced by Levene and Poulovassilis (1990). They defined the model and a declarative logic-based language structured as a sequence of instructions (hypernode programs), used for querying and updating hypernodes. Later, Poulovassilis and Levene (1994) included the notion of schema and type checking, introduced via the idea of types (primitive and complex), that were also represented by nested graphs (See an example in Fig. 1.4). They also included a rule-based query language called *Hyperlog*, which can support both querying and browsing using logical rules as well as database updates, and is intractable in the general case. In the third version of the model, Levene and Loizou (1995) discussed a set of constraints (entity, referential and semantic) over hypernode databases. Additionally, they proposed another query and update language called HNQL, which uses compound statements to produce HNQL programs.

Summarizing, the main features of the Hypernode model are: a nested graph structure that is simple and formal; the ability to model arbitrary complex objects in a straightforward manner; underlying data structure of an object-oriented data model; enhancement of the usability of a complex objects database system via a graph-based user interface.

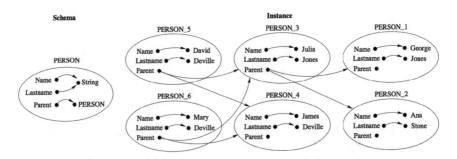

Fig. 1.4 Hypernode Model. The schema (left) defines a *person* as a complex object with the properties *name* and *lastname* of type string, and *parent* of type person (recursively defined). The instance (on the right) shows the relations in the genealogy among different instances of person

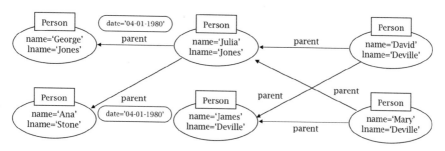

Fig. 1.5 Property graph data model. The main characteristic of this model is the occurrence of properties in nodes and edges. Each property is represented as a pair property-name = "property-value"

1.3.4 The Property Graph Model

A *property graph* is a directed, labeled, attributed multigraph. That is, a graph where the edges are directed, both nodes and edges are labeled and can have any number of properties (or attributes), and there can be multiple edges between any two nodes (Rodriguez and Neubauer 2010). Properties are key/value pairs that represent metadata for nodes and edges. In practice, each node of a property graph has an identifier (unique within the graph) and zero or more labels. Node labels could be associated to node typing in order to provide schema-based restrictions. Additionally, each (directed) edge has a unique identifier and one or more labels. Figure 1.5 shows an example of property graph.

Property graphs are used extensively in computing as they are more expressive[1] than the simplified mathematical objects studied in theory. In fact, the property graph model can express other types of graph models by simply abandoning or adding particular features or components (Rodriguez and Neubauer 2010).

There is no standard query language for property graphs although some proposals are available. Blueprints (2018) was one of the first libraries created for the property graph data model. Blueprints is analogous to JDBC, but for graph databases. Gremlin (2018) is a functional graph query language that allows to express complex graph traversals and mutation operations over property graphs. Neo4j (2018) provides Cypher (2018), a declarative query language for property graphs. The syntax of Cypher, very similar to SQL via expressions match-where-return, allows to easily express graph patterns and path queries. PGQL (van Rest et al. 2013), a graph query language designed by Oracle researchers, is closely aligned to SQL and supports powerful regular path expressions. G-CORE (Angles et al. 2018) is a recent proposal that integrates the main and relevant features provided by old and current graph query languages.

---

[1]Note that the expressiveness of a model is defined by ease of use, not by the limits of what can be modeled.

1.3.5 Web Data Graphs: The RDF Model

The Resource Description Framework (RDF) (Klyne and Carroll 2004) is a recommendation of the W3C designed originally to represent metadata. One of the main advantages (features) of the RDF model is its ability to interconnect resources in an extensible way using graph-like structure for data.

One of the main advantages of RDF is its dual nature. In fact, there are two possible readings of the model. From a knowledge representation perspective, an atomic RDF expression is triple consisting of a subject (the resource being described), a predicate (the property) and an object (the property value). Each triple represents a logical statement of a relationship between the subject and the object, and one could enhance this basic logic by adding rules and ontologies over it (e.g., RDFS and OWL) A general RDF expression is a set of such triples called an RDF Graph (see example in Fig. 1.6), which can be intuitively considered as a semantic network. From the second perspective, the RDF model is the most general representation of a graph, where edges are also considered nodes. In this sense, formally it is not a traditional graph (Hayes and Gutierrez 2004). This allows self-references, reification (i.e., making statements over statements), and that it is essentially self-contained. The drawback of all these features is the complexity that comes with this generalization, particularly for efficient implementation.

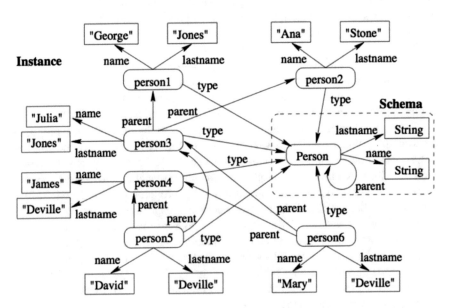

Fig. 1.6 RDF data model. Note that schema and instance are mixed together. The edges labeled *type* disconnect the instance from the schema. The instance is built by the subgraphs obtained by instantiating the nodes of the schema, and establishing the corresponding parent edges between these subgraphs

SPARQL (Prud'hommeaux and Seaborne 2008) is the standard query language for RDF. It is able to express complex graph patterns by means of a collection of triple patterns whose solutions can be combined and restricted by using several operators (i.e., AND, UNION, OPTIONAL and FILTER). The latest version of the language, SPARQL 1.1 (Harris and Seaborne 2013), includes explicit operators to express negation of graph patterns, arbitrary length path matching (i.e., reachability), aggregate operators (e.g., COUNT), subqueries and query federation.

1.4 Querying Graph Databases

Data manipulation and querying in graph data management is expressed by graph operations or graph transformations whose main primitives are based on graph features like neighborhoods, graph patterns and paths. Another approach is to express all queries using a few powerful graph manipulation primitives enclosed by a graph query language.

This section contains a brief overview of the research on querying graph databases. First, we present a broad classification of queries studied in the context of graph databases, including a description of their characteristics (e.g., complexity and expressiveness). After that, we present a review of graph query languages, including short descriptions of some proposals we consider representative of the area.

1.4.1 Classification of Graph Queries

In this section, we present a broad classification of queries that have been largely studied in graph theory and can be considered essential for graph databases. We grouped them in adjacency, pattern matching, reachability and analytical queries.

To fix notations, let us represent a graph database as a single-labeled directed multigraph. Specifically, a tuple $G = (N, E, L, \delta, \lambda_N, \lambda_E)$, where N is a finite set of nodes, E is a finite set of edges, L is a finite set of labels, $\delta : E \rightarrow N^2$ is the edge function that associates edges with pairs of nodes, $\lambda_N : N \rightarrow L$ is the node labeling function, and $\lambda_E : E \rightarrow L$ is the edge labeling function. An edge $e = (n, n') \in E$ will be represented as a triple (v, w, v') where $v = \lambda_N(n)$, $w = \lambda_E(e)$ and $v' = \lambda_N(n')$. Nodes and edges will usually be referenced by using their labels. Additionally, a path ρ in G is a sequence of edges (v_0, w_0, v_1), (v_1, w_1, v_2), ..., (v_{m-1}, w_{m-1}, v_m), where v_0 and v_m are the source and target nodes of the path, respectively. The label of ρ is the sequence of labels $w_0, w_1, \ldots, w_{m-1}$.

1.4.1.1 Adjacency Queries

The primary notion in this type of queries is node/edge adjacency. Two nodes are adjacent (or neighbors) when there is an edge between them. Similarly, two edges are adjacent when they share a common node. Examples of adjacency queries are: "return the neighbors of a node v" or "check whether nodes v and v' are adjacent." In spite of their simplicity, to compute efficiently adjacency queries could be a challenge for big sparse graphs (Kowalik 2007).

The basic notion of adjacency can be extended to define more complex *neighborhood queries*. For instance, the k-neighborhood of a root node v is the set of all nodes that are reachable from v via a path of k edges, that is, the length of the path is no more than k (Papadopoulos and Manolopoulos 2005). Similarly, the k-hops of v returns all the nodes that are at a distance of k edges from v. Note that a k-neighborhood query can be expressed as a composition of j-hops queries using set union as 1-hops $\cup \cdots \cup k$-hops (Dominguez-Sal et al. 2010a).

Several applications can benefit from adjacency queries, in particular those where the notion of influence is an important concern. For instance, in information retrieval adjacency queries are used for web ranking using hubs and authorities (Chang and Chen 1998). In recommendation systems, they are used to obtain users with similar interests (Dominguez-Sal et al. 2010a). In social networks, they can be used to validate the well-known six-degrees-of-separation theory.

1.4.1.2 Pattern Matching Queries

The basic notion of graph pattern matching consists in finding the set of subgraphs of a database graph that "match" a given graph pattern. A basic graph pattern is usually defined as a small graph where some nodes and edges can be labeled with variables. The purpose of the variables is to indicate unknown data and more importantly, to define the output of the query (i.e., variables will be "filled" with solution values). For instance, the expression $(John, friend, ?y)$, $(John, friend, ?z)$, $(?y, friend, ?z)$ represents a graph pattern where $?x$ and $?y$ are variables. The result or interpretation of this graph pattern could be "the pairs of friends of John who are also friends."

Graph pattern matching is typically defined in terms of subgraph isomorphism, that is, to find all subgraphs of a database G that are isomorphic to a graph pattern P. Hence, pattern matching deals with two problems: the graph isomorphism problem that has a unknown computational complexity, and the subgraph isomorphism problem that is an NP-complete problem (Gallagher 2006).

Graph pattern matching is easily identifiable in many application domains. For instance, graph patterns are fundamental within the pattern recognition field (Conte et al. 2004). In social network analysis, it is used to identify communities and social positions (Fan 2012). In protein interaction networks, researchers are interested in patterns that determine proteins with similar functions (Tian et al. 2007).

There are a number of variations on the basic notion of pattern matching:

- *Graph patterns with structural extension or restrictions.* A basic graph pattern has been defined as a simple structure containing nodes, edges and variables; however, this notion can be extended or restricted depending on the graph data model. For instance, if the database is a property graph then a graph pattern should support conditions over such properties.
- *Complex graph patterns.* In some cases, a collection of basic graph patterns can be combined via specific operators (e.g., union, optional and difference) to conform complex graph patterns. The semantics of these graph patterns can be defined in terms of an algebra of graph patterns.
- *Semantic matching.* It consists in matching graphs based on specific interpretations (i.e., semantics) given to nodes and edges. Such interpretations can be defined via semantic rules (e.g., an ontology).
- *Inexact matching.* In this case the graph pattern matching algorithm returns a ranked list of the most similar matches (instead of the original exact matching). These algorithms employ a cost function to measure the similarity of the graphs and error correction techniques to deal with noisy data.
- *Approximate matching.* This variation concerns the use of algorithms that find approximate solutions to the pattern matching problem, that is, they offer polynomial time complexity but are not guaranteed to find a solution. In case of exact matching the algorithm will return some solutions, but not all matches. For inexact matching, a close solution will be returned, but not the closest.

Very related to graph pattern matching is the area of *graph mining* (Aggarwal and Wang 2010). This area includes the problems of frequent pattern mining, clustering and classification. For instance, the goal of frequent pattern mining is the discovery of common patterns, that is, to find subgraphs that occur frequently in the entire database graph. The problem of computing frequent subgraphs is particularly challenging and computationally intensive, as it needs to compute graph and subgraph isomorphisms. The discovery of patterns can be useful for many application domains, including finding strongly connected groups in social networks and finding frequent molecular structures in biological databases.

1.4.1.3 Reachability Queries (Connectivity)

One of the most characteristic problems in graph databases is to compute reachability of information. In general terms, the problem of reachability tests whether two given nodes are connected by a path. Reachability queries have been addressed in traditional database models, in particular for querying relational and semistructured databases (Agrawal and Jagadish 1987; Abiteboul and Vianu 1999). Yannakakis (1990) surveyed a set of path problems relevant to the database area, including computing transitive closures, recursive queries and the complexity of path searching.

In the context of graph databases, reachability queries are usually modeled as path or traversal problems characterized by allowing restrictions over nodes and edges. Cruz et al. (1987) introduced the notion of Regular Path Query (RPQ) as a way of expressing reachability queries. The basic structure of a regular path query is an expression $(?x, \tau, ?y)$, where $?x$ and $?y$ are variables, and τ is a regular expression. The goal of this RPQ is to find all pairs of nodes $(?x, ?y)$ connected by a path such that the concatenation of the labels along the path satisfies τ. Note that variables $?x$ and $?y$ can be replaced by node labels (i.e., data values) in order to define specific source and target nodes, respectively. For instance, the path query $(John, friend^+, ?z)$ returns the people $?z$ that can be reached from "John" by following "friend" edges.

The complex nature of path problems is such that their computations often require a search over a sizable data space. The complexity of regular path queries was initially studied by Mendelzon and Wood (1995) in terms of computing simple paths (i.e., paths with no repeated nodes). Specifically, the problem of finding all pairs of nodes connected by a simple path satisfying a given regular expression was shown to be NP-complete in the size of the graph. Due to the high computational complexity of RPQs under simple path semantics, researchers proposed a semantics based on arbitrary paths. This semantics leads to tractable combined complexity for RPQs and tractable data complexity for a family of expressive languages. See the work of Barceló Baeza (2013) for a complete review about these issues.

Reachability queries are present in multiple application domains: in semistructured data they are used to query XML documents using XPath (Abiteboul and Vianu 1999); in social networks they allow to discover people with common interests (Fan 2012); and in biological networks they allow to find specific biochemical pathways between distant nodes (Tian et al. 2007). Additionally, reachability queries are the basis for other real-life graph queries. Maybe the most important is the *shortest-path distance* (also called the geodesic distance). For instance, in a road network it is fundamental to calculate the minimum distance between two locations (Zhu et al. 2013).

1.4.1.4 Analytical Queries

The queries of this type do not consult the graph structure; instead they are oriented to measure quantitatively and usually in aggregate form topological features of the database graph. Analytical queries can be supported via special operators that allow to summarize the query results, or by ad hoc functions hiding complex algorithms.

Summarization queries can be expressed in a query language by using the so-called aggregate operators (e.g., average, count, maximum, etc.). These operators can be used to calculate the order of the graph (i.e., the number of nodes), the degree of a node (i.e., the number of neighbors of the node), the minimum/maximum/average degree in the graph, the length of a path (i.e., the number of edges in the path), the distance between nodes (i.e., the length of a shortest path between the nodes), among other "simple" analytical queries.

Complex analytical queries are related to important algorithms for graph analysis and mining (see the work of Aggarwal and Wang (2010) for an extensive review). Examples of such graph algorithms are:

- *Characteristic path length.* It is the average shortest path length in a network. It measures the average degree of separation between the nodes.
- *Connected components.* It is an algorithm for extracting groups of nodes that can reach each other via graph edges.
- *Community detection.* This algorithm deals with the discovery of groups whose constituent nodes form more relationships within the group than with nodes outside the group.
- *Clustering coefficient.* The clustering coefficient of a node is the probability that the neighbors of the node are also connected to each other. The average clustering coefficient of the whole graph is the average of the clustering coefficients of all individual nodes.
- *PageRank* This algorithm, created in the context of web searching, models the behavior of an idealized random web surfer. The PageRank score of a web page represents the probability that the random web surfer chooses to view the web page. This algorithm can be an effective method to measure the relative importance of nodes in a data graph.

Complex analytical queries are the speciality of graph-processing frameworks due to their facilities for implementing and running complex algorithms over large graphs. More details about these type of queries can be found in articles about graph-processing frameworks (Guo et al. 2014; Zhao et al. 2014).

1.4.2 A Short Review of Graph Query Languages

In the literature of graph data management there is substantial work on graph query languages (GQLs). A review of GQLs proposed during the first wave of graph databases was presented by Angles and Gutierrez (2008). Based on this, Wood (2012) studied several GQLs focusing on their expressive power and computational complexity. A review and comparison of practical query languages provided by graph databases (available at the time) was presented by Angles (2012). Barceló Baeza (2013) studied the problem of querying graph databases, in particular the expressiveness and complexity of several navigational query languages. Recently, Angles et al. (2017) presented a survey of the foundational features underlying modern graph query languages.

Due to space constraint, we will not present a complete review of graph query languages. Instead, we describe some of the languages we consider relevant and useful to show the developments in the area. Moreover, we restrict our review to "pure" GQLs, that is, those languages specifically designed to work with graph data models. Figure 1.7 presents this subset of languages in chronological order.

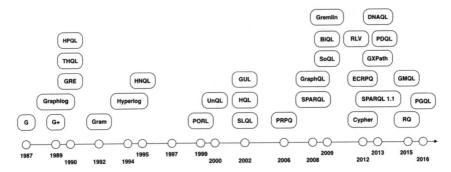

Fig. 1.7 Evolution of graph query languages: G (Cruz et al. 1987), G+ (Cruz et al. 1989), Graphlog (Consens and Mendelzon 1990), GRE (Wood 1990), THQL (Watters and Shepherd 1990), HPQL (Levene and Poulovassilis 1990), HML (Levene and Poulovassilis 1991), Gram (Amann and Scholl 1992), Hyperlog (Poulovassilis and Levene 1994), HNQL (Levene and Loizou 1995), HQL (Theodoratos 2002), PRPQ (Liu and Stoller 2006), SPARQL (Prud'hommeaux and Seaborne 2008), GraphQL (He and Singh 2008), Gremlin (Rodriguez 2015), Cypher (2018), SPARQL 1.1 (Harris and Seaborne 2013), PGQL (van Rest et al. 2013), PDQL (Angles et al. 2013) and G-CORE (Angles et al. 2018)

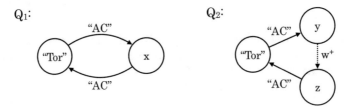

Fig. 1.8 Example of a graphical graph query expressed in the G query language

Depending on their inherent data model, the query languages can be grouped in: languages for edge-labeled graphs (G, G+, Graphlog, GRE, Gram and PDQL), languages for hypergraphs (HML, THQL and HQL), languages for nested graphs (HPQL, Hyperlog and HNQL), languages for property graphs (PRPQ, GraphQL, Gremlin, Cypher, PGQL and G-CORE), and RDF query languages (SPARQL and SPARQL 1.1). Next, we present a brief description of some of them.

Cruz et al. (1987) proposed G, a query language that introduced the notion of graphical query as a set of query graphs. A query graph (pattern) is a labeled, directed multigraph in which the node labels may be either variables or constants, and the edge labels can be regular expressions combining variables and constants. The result of a graphical query Q with respect to a graph database G is the union of all query graphs of Q which match subgraphs of G.

Figure 1.8 presents an example of a graphical query containing two query graphs, Q_1 and Q_2. This query finds the first and last cities visited in all round-trips from Toronto ("Tor"), in which the first and last flights are with Air Canada ("AC") and all other flights (if any) are with the same airline. Note that the last condition is expressed by the edge labeled with regular expression w^+. Thanks to the inclusion of regular expressions, G is able to express recursive queries more general than transitive closure. However, the evaluation of queries in G is of high computational complexity due to its semantics based on simple paths.

G evolved into a more powerful language called G+ (Cruz et al. 1989). The notion of graphical query proposed by G is extended in G+ to define a summary graph that represents how to restructure the answer obtained by the query graphs. Additionally, G+ allows to express aggregate functions over paths and sets of paths (i.e., it allows to compute the size of the shortest path).

GraphLog (Consens and Mendelzon 1989) is a query language that extends G+ by adding negation and unifying the concept of a query graph. A query is now a single graph pattern containing one distinguished edge that corresponds to the restructured edge of the summary graph in G+. The effect of a GraphLog query is to find all instances of the pattern that occur in the database graph and for each one of them define a virtual link represented by the distinguished edge. Consens and Mendelzon (1990) have shown that the expressive power of GraphLog is equivalent to three well-known query classes: stratified linear Datalog programs, queries computable in nondeterministic logarithmic space, and queries expressible with a transitive closure operator plus first-order logic. Based on this, the GraphLog's authors argued that the language is able to express "real life" recursive queries.

Gram (Amann and Scholl 1992) is a query language based on walks[2] and hyperwalks. Assuming that T is the union of node and edge types in the database graph, a walk expression is a regular expression over T without alternation (union), whose language contains only alternating sequences of node and edge types. A hyperwalk is a set of walk expressions connected by at least one node type. Assuming a database graph containing travel agency data, the expression JOURNEY first (STOP next)* + STOP in CITY is a hyperwalk containing two walk expressions connected by the node type STOP. Hence, the above hyperwalk describes the walks going from a node (of type) JOURNEY to one of its nodes (of type) STOP in a CITY. The set of walks in the database satisfying a hyperwalk expression r is called the instance of r and is denoted by $I(r)$. Based on these notions, Gram defines a hyperwalk algebra with operations closed under the set of hyperwalks (e.g., projection, selection, join and set operations). For example, the algebra expression

$$\pi_{JOURNEY}(\sigma_{Munich(CITY)}I(\text{JOURNEY first(STOP next)* STOP in CITY}))$$

computes all journeys that traverse Munich.

Although less popular, there are also languages for manipulating and querying hypergraphs and hypernodes (nested graphs). For instance, GROOVY (Levene and Poulovassilis 1991) introduced a Hypergraph Manipulation Language (HML) for querying and updating labeled hypergraphs, which defines basic operators for

---

[2]In graph theory, a walk is an alternating sequence of nodes and connecting edges, which begins and ends with a node, and where any node and any edge can be visited any number of times.

manipulation (addition and deletion) and querying of hypergraphs and hyperedges. On the other side, Levene and Poulovassilis (1990) defined a logic-based query and update language for hypernodes where a query is expressed as a hypernode program consisting of a set of hypernode rules.

GraphQL (He and Singh 2008) is a graph query language for property graphs, which is based on the use of formal grammars for composing and manipulating graph structures. A graph grammar is a finite set of graph motifs where a graph motif can be either a simple graph or composed of other graph motifs by means of concatenation, disjunction and repetition. For instance, consider the following graph grammar containing three graph motifs:

graph G_1 { **node** v_1, v_2; **edge** $e_1(v_1,v_2)$; }
graph G_2 { **node** v_2, v_3; **edge** $e_2(v_2,v_3)$; }
graph G_3 { **graph** G_1 as X; **graph** G_2 as Y; **edge** $e_3(X.v_2, Y.v_2)$ }.

The graph motifs G_1 and G_2 are simple, whereas G_3 is a complex graph motif that concatenates the graph motifs G_1 and G_2 via the edge e_3 and the common node v_2. The language of a graph grammar is the set of all the graphs derivable from graph motifs of that grammar. The query language is based on graph patterns consisting of a graph motif plus a predicate on attributes of the motif. A predicate is a combination of boolean or arithmetic comparison expressions. For instance, the expression

graph P { **node** v_1, v_2; **edge** $e_1(v_1,v_2)$ }
where v_1.name="A" and v_2.year > 2000;

describes a graph pattern where two nodes v_1, v_2 must be connected by an edge e_1, and the nodes must satisfy the conditions following the **where** clause.

Note that most of the languages described above are more theoretical than practical. Cypher (2018) is a declarative language for querying property graphs implemented by the Neo4j graph database. The most basic query in Cypher consists of an expression containing clauses START, MATCH and RETURN. For example, assuming a friendship graph, the following query returns the name of the friends of the persons named "John":

```
START x=node:person(name="John")
MATCH (x)-[:friend]->(y)
RETURN y.name
```

The START clause specifies one or more starting points (nodes or edges) in the database graph. The MATCH clause contains the graph pattern of the query. The RETURN clause specifies which nodes, edges and properties in the matched data will be returned by the query.

Cypher is able to express some types of reachability queries via path expressions. For instance, the expression p = (a)-[:knows*]->(b) computes the paths from node (a) to node (b), following only knows outgoing edges, and maintains the solution in the path variable p. Additionally, there exist built-in functions to calculate specific operations on nodes, edges, attributes and paths. For instance, complementing the above path expression, the function shortestPath(p) returns the shortest path between nodes (a) and (b).

SPARQL (Prud'hommeaux and Seaborne 2008) is the standard query language for the RDF data model. A typical query in SPARQL follows the traditional SELECT-FROM-WHERE structure where the FROM clause indicates the data sources, the WHERE clause contains a graph pattern, and the SELECT clause defines the output of the query (e.g., resulting variables). The simplest graph pattern, called a triple pattern, is an expression of the form subject-predicate-object where identifiers (i.e., URIs), values (RDF Literals) or variables (e.g., ?X) can be used to represent a node-edge-node pattern. A complex graph pattern is a collection of triple patterns whose solutions can be combined and restricted by using operators like AND, UNION, OPTIONAL and FILTER. For instance, the following query returns the names of persons described in the given data source (i.e., an RDF graph):

```
SELECT ?N
FROM <http://example.org/data.rdf>
WHERE { ?X rdf:type voc:Person . ?X voc:name ?N }
```

The latest version of the language, SPARQL 1.1 (Harris and Seaborne 2013), includes novel features like negation of graph patterns, arbitrary length path matching (i.e., reachability), aggregate operators (e.g., COUNT), subqueries and query federation.

Although a GQL is normally related to a graph database model, this relation is not exclusive. For instance, several object-oriented data models defined graph-based languages to manipulate the objects in the database (e.g., GraphDB and G-Log), or to represent database transformations (e.g., GOOD and GUL). A similar situation occurred for semistructured data models when graph-oriented operations were used to navigate the tree-based data (e.g., Lorel and UnQL). Additionally, several graph-based query languages have been designed for specific applications domains, in particular those related to complex networks, for instance social networks (Ronen and Shmueli 2009), biological networks (Brijder et al. 2013), bibliographical networks (Dries et al. 2009), the web (Dries et al. 2009) and the Semantic Web (Harris and Seaborne 2013).

1.5 Graph Data Management Systems

The systems for graph data management can be classified into two main categories: graph databases and graph-processing frameworks. Although the problems addressed for both groups are similar, they provide two different approaches for storing and querying graph data, with their own advantages and disadvantages.

Graph databases aim at persistent management of graph data, allowing to transactionally store and access graph data on a persistent medium. In this sense, these provide efficient single-node solutions with limited scalability. On the other hand, graph-processing frameworks aim to provide batch processing and analysis of large graphs often in a distributed environment with multiple machines. These solutions usually process the graph in memory, but different parts of the graph are managed by distinct, distributed nodes.

Closely related to graph databases are the systems for managing RDF data. These systems, called RDF Triple Stores or RDF database systems, are specifically designed to store collections of RDF triples, to support the standard SPARQL query language, and possibly to allow some kind of inference via semantic rules. Although Triple Stores are based on the RDF graph data model, they are specialized databases with their own characteristics. Therefore, we will study them separately.

Next we present a review of current systems in the above categories, including a short description of each of them.

1.5.1 Graph Database Systems

A graph database system (GDBS)—or just graph database—is a system specifically designed for managing graph-like data following the basic principles of database systems, that is, persistent data storage, physical/logical data independence, data integrity and consistency. The research on graph databases has a long history, at least since the 1980s. Although the first of these were primarily theoretical proposals (with emphasis on graph database models), it is only recently that several technological developments (e.g., powerful hardware to store and process graphs) have made it possible to have practical systems.

The current "market" of graph databases includes systems providing most of the major components in database management systems, including: storage engine (with innate support for graph structures), database languages (for data definition, manipulation and querying), indexes and query optimizer, transactions and concurrency controllers, and external interfaces (user interface or API) for system management.

Considering their internal implementation, we classify graph databases in two types: native and nonnative graph databases. Native graph databases (see Table 1.1) implement ad hoc data structures and indexes for storing and querying graphs. Nonnative graph databases (see Table 1.2) make use of other database systems to store graph data and implement query interfaces to execute graph queries over the back-end system. Some of these systems are described below.

Table 1.1 List of native graph database systems

System	URL
Amazon Neptune	https://aws.amazon.com/neptune/
AllegroGraph	http://www.franz.com/agraph/allegrograph/
GraphBase	https://graphbase.ai/
GraphChi	https://github.com/GraphChi
HyperGraphDB	http://www.hypergraphdb.org/
InfiniteGraph	http://www.objectivity.com/products/infinitegraph/
InfoGrid	http://infogrid.org/
Neo4j	http://neo4j.com/
Sparksee/DEX	http://www.sparsity-technologies.com/
TigerGraph	https://www.tigergraph.com/

Table 1.2 List of nonnative graph database systems

System	URL
ArangoDB	http://www.arangodb.org
FlockDB	https://github.com/twitter/flockdb/
JanusGraph	http://janusgraph.org/
Microsoft Cosmos DB	https://docs.microsoft.com/en-us/azure/cosmos-db
OQGraph	https://mariadb.com/kb/en/mariadb/oqgraph-storage-engine/
Oracle spatial and graph	http://www.oracle.com/technetwork/database/options/spatialandgraph/
OrientDB	http://orientdb.com
Titan	http://thinkaurelius.github.io/titan/
VelocityGraph	https://velocitydb.com/VelocityGraph.aspx

AllegroGraph is one of the precursors in the current generation of graph databases. It combines efficient memory utilization and disk-based storage. Some of the most interesting features of AllegroGraph is its support for Lisp and Prolog interfaces for querying the database. Although it was born as a graph database, its current development is oriented to meet the Semantic Web standards. Additionally, AllegroGraph provides special features for GeoTemporal Reasoning and Social Network Analysis.

Neo4j is a native graph database that supports transactional applications and graph analytics. Neo4j is based on a network-oriented model where relations are first-class objects. It is fully written in Java and implements an object-oriented API, a native disk-based storage manager for graphs, and a framework for graph traversals. Cypher is the declarative graph query language provided by Neo4j.

Sparksee (formerly DEX) is a native graph database for persistent storage of property graphs. Its implementation is based on bitmaps and other secondary structures, and provides libraries (APIs) in several languages for implementing graph queries. Sparksee is being used in social, bibliographical and biological networks analysis, media analysis, fraud detection and business intelligence applications of indoor positioning systems.

HyperGraphDB is a system that implements the hypergraph data model (i.e., edges are extended to connect more than two nodes). This model allows a natural representation in higher-order relations, and is particularly useful for modeling data of areas like knowledge representation, artificial intelligence and bio-informatics. Hypergraph stores the graph information in the form of key/value pairs that are stored on BerkeleyDB.

InfiniteGraph is a database oriented to support large-scale graphs in a distributed environment. It aims the efficient traversal of relations across massive and distributed data stores. Its focus of attention is to extend business, social and government intelligence with graph analysis.

There are several papers comparing the features (Dominguez-Sal et al. 2010b; Angles 2012; McColl et al. 2013) and performance (Vicknair et al. 2010; Ciglan et al. 2012; Jouili and Vansteenberghe 2013) of graph databases. Additionally, industrial benchmarking results for graph databases are provided by the Linked Data Benchmark Council through the Social Network Benchmark (Erling et al. 2015).

1.5.2 Graph-Processing Frameworks

In addition to graph databases, a number of graph-processing frameworks have been proposed to address the needs of processing complex and large-scale graph datasets. These frameworks are characterized by in-memory batch processing and the use of distributed and parallel-processing strategies. Note that distributed systems with more computing and memory resources are able to process large-scale graphs, but they can be less efficient than single-node platforms when specific graph queries are executed.

On the one hand, generic data processing systems such as Hadoop, YARN, Stratosphere and Pegasus have been adapted for graph processing due to their facilities for batch data processing. Most of these systems are based on the MapReduce programming model and implemented on top of the Hadoop platform, the open-source version of MapReduce. By exploiting data-parallelism, these systems are highly scalable and support a range of fault-tolerance strategies. Though these systems improve the performance of iterative queries, users still need to "think of" their analytical graph queries as MapReduce jobs. It is important to note that implementing graph algorithms in these data-parallel abstractions can be challenging (Xin et al. 2013). Additionally, these systems cannot take advantage of the characteristics of graph-structured data and often result in complex job chains and excessive data movement when implementing iterative graph algorithms (Zhao et al. 2014).

On the other hand, graph-specific platforms (see Table 1.3) provide different programming interfaces for expressing graph analytic algorithms. These platforms, also called *offline graph analytic systems*, perform an iterative, batch processing over the entire graph dataset until the computation satisfies a fixed-point or stopping criterion. Therefore, these systems are particularly designed for computing graph

Table 1.3 List of graph-processing frameworks

System	URL
Apache Giraph	http://giraph.apache.org
BLADYG	https://members.loria.fr/saridhi/files/software/bladyg/
GPS	http://infolab.stanford.edu/gps/
GraphLab	https://turi.com/
GraphX	https://spark.apache.org/graphx/
Ligra	http://jshun.github.io/ligra/docs/introduction.html
Microsoft GraphEngine	https://www.graphengine.io/
PowerGraph	https://github.com/jegonzal/PowerGraph

algorithms that require iterative, batch processing, for example, PageRank, recursive relational queries, clustering, social network analysis, machine learning and data mining algorithms (Khan and Elnikety 2014). Next, we briefly describe some of these systems.

Pregel (Malewicz et al. 2010) is an API designed by Google for writing algorithms that process graph data. Pregel is a node-centric programming abstraction that adapts the Bulk Synchronous Parallel (BSP) model, which was developed to address the problem of parallelizing jobs across multiple workers for scalability. The fundamental computing paradigm of Pregel, called "think like a node," defines that graph computations are specified in terms of what each node has to compute; edges are communication channels for transmitting computation results from one node to another, and do not participate in the computation. To avoid communication overheads, Pregel preserves data locality by ensuring computation is performed on locally stored data.

Apache Giraph is an open-source implementation of Google Pregel. Giraph runs workers as map-only jobs on Hadoop and uses HDFS for data input and output. Giraph also uses Apache ZooKeeper for coordination, checkpointing and failure recovery schemes. Giraph has incorporated several optimizations, has a rapidly growing user base, and has been scaled by Facebook to graphs with a trillion edges. Giraph is executed in-memory, which can speed-up job execution, but, for large amounts of messages or big datasets, can also lead to crashes due to lack of memory.

GraphLab (Low et al. 2012) is an open-source, graph-specific distributed computation platform implemented in C++. GraphLab uses the GAS decomposition (Gather, Apply, Scatter), which looks similar to, but is fundamentally different from, the BSP model. In the GAS model, a node accumulates information about its neighborhood in the Gather phase, applies the accumulated value in the Apply phase, and updates its adjacent nodes and edges and activates its neighboring nodes in the Scatter phase. Another key difference is that GraphLab partitions graphs using vertex cuts rather than edge cuts. Consequently, each edge is assigned to a unique machine, while nodes are replicated in the caches of remote machines. Besides graph processing, it also supports various machine learning algorithms.

Apache GraphX is an API for graphs and graph-parallel computation implemented on top of Apache Spark (a general platform for big data processing). GraphX unifies ETL, exploratory analysis, and iterative graph computation within a single system. GraphX extends Spark with graphs based on Sparks Resilient Distributed Datasets (RDDs). It allows to view the same data as both graphs and collections, transform and join graphs with RDDs efficiently, and write custom iterative graph algorithms.

There is an increasing body of work comparing graph-processing frameworks. For instance, the first evaluation study of modern big data frameworks, including Map-Reduce, Stratosphere, Hama, Giraph and Graphlab was presented by Elser and Montresor (2013). Guo et al. (2014) presented a benchmarking suite for graph-processing platforms. The suite was used to evaluate the performance of Hadoop, YARN, Stratosphere, Giraph, GraphLab and Neo4j. Zhao et al. (2014) presented a comparison study on parallel-processing systems, including Giraph, GPS and

GraphLab. Han et al. (2014) presented a comparison considering optimizations and several metrics done among Giraph, GPS, Mizan and GraphLab. LDBC Graphalytics (Iosup et al. 2016) is an industrial-grade benchmark for large-scale graph analysis on parallel and distributed platforms.

1.5.3 RDF Database Systems

An RDF database (also called *Triple Store*) is a specialized graph database for managing RDF data. RDF defines a data model based on expressions of the form subject-predicate-object (SPO) called RDF triples. Therefore, an RDF dataset is composed of a large collection of RDF triples that implicitly form a graph. SPARQL is the standard query language for RDF databases. It is a declarative language that allows to express several types of graph patterns. Its most recent version (SPARQL 1.1) supports advanced features like property paths, aggregate functions and subqueries. Table 1.4 presents a list of RDF database systems. These systems can be classified in three types: native RDF stores, relational-based RDF stores and graph-based RDF stores.

A native RDF store is designed and optimized for the storage and retrieval of RDF triples. The main challenge in this type of system is to all six permutation indexes on the RDF data in order to provide efficient query processing for all possible access patterns (Yuan et al. 2013; Atre et al. 2010). Examples of RDF stores are Jena, RDF-3X (Yuan et al. 2013), 4store (Harris et al. 2009), TripleBit (Yuan et al. 2013), HexaStore (Weiss et al. 2008), GraphDB and BrightstarDB.

There are several approaches for managing RDF data with a relational database. A *triple table* refers to the approach of storing RDF data in a three-column table with each row representing an SPO statement (Neumann and Weikum 2010). A second approach is to store RDF data in a *property table* (Abadi et al. 2007; Chong et al. 2005) with subject as the first column and the list of distinct predicates as the

Table 1.4 List of RDF database systems

System	URL
Blazegraph	https://www.blazegraph.com/
BrightstarDB	http://brightstardb.com
GraphDB	https://ontotext.com/products/graphdb/
gstore	http://www.icst.pku.edu.cn/intro/leizou/projects/gStore.htm
Jena	https://jena.apache.org/
RDF-3X	https://code.google.com/archive/p/rdf3x/
Stardog	http://stardog.com
TripleBit	http://grid.hust.edu.cn/triplebit/
Virtuoso	http://virtuoso.openlinksw.com/
3store	http://sourceforge.net/projects/threestore/

remaining columns. RDF data can also be stored by using multiple two-column tables, one for each unique predicate. The first column is for subject, whereas the other column is for object. This method, called *column store with vertical partitioning* (Abadi et al. 2007), can be implemented over row-oriented or column-oriented database systems. A third mechanism, called *entity-oriented*, treats the columns of a relation as flexible storage locations that are not preassigned to any predicate, but predicates are assigned to them dynamically, during insertion. The assignment ensures that a predicate is always assigned to the same column or more generally the same set of columns (Bornea et al. 2013). Some examples of relational-based RDF stores are Virtuoso and 3store.

A graph-based approach focuses on storing RDF data as a graph (Zou et al. 2014; Zeng et al. 2013; Morari et al. 2014). In this case, the RDF triples must be modeled as classical graph nodes and edges, and the SPARQL queries must be transformed into graph queries. Among the systems of this type we can mention gstore, Stardog, TripleBit and Blazegraph.

There are several works comparing RDF databases (see, for example Schmidt et al. 2008; Stegmaier et al. 2009; Faye et al. 2012; Cudré-Mauroux et al. 2013). The Semantic Publishing Benchmark is a proposal of standard benchmark for evaluating RDF database systems (Kotsev et al. 2016).

1.6 Conclusions

Graph data management is currently an ongoing and fast-developing area with manifold application domains and increasing industrial interest. Today, there is a broad landscape of systems and models for graph database management and deep theoretical research on data models, query languages and algorithms that address the challenges of the area. Slowly and increasingly, academia and industry are reaching consensus on several of its features, such as data models, data formats, query languages and benchmarks. All this is the consequence of having growing amounts of datasets for which graphs are their natural model. In this regard, we are living exciting times for graph data management.

Many challenges remain, though. Among the most important are the standardization of data formats and query languages; the integration of graph systems with other models, particularly the relational one; the deepening of the understanding of the most novel features that graphs bring to the world of data, like paths, connectivity and such; and the presentation and visualization of graph data. In this regard, the creation of initiatives like The Linked Data Benchmark Council (http://www.ldbcouncil.org/), Open Cypher (https://www.opencypher.org/), Linked Open Data (http://linkeddata.org/) and several company developments (e.g., Amazon Neptune, Microsoft Azure Cosmos DB, Oracle Spatial and Graph) are relevant to support and spur the development of graph data management.

Acknowledgements R. Angles and C. Gutierrez were supported by the Millennium Nucleus Center for Semantic Web Research under grant NC120004.

References

Abadi DJ, Marcus A, Madden SR, Hollenbach K (2007) Scalable semantic web data management using vertical partitioning. In: Proceedings of the international conference on very large data bases (VLDB), pp 411–422

Abiteboul S, Vianu V (1999) Regular path queries with constraints. J Comput Syst Sci 58:428–452

Abiteboul S, Quass D, McHugh J, Widom J, Wiener JL (1997) The Lorel query language for semistructured data. Int J Digit Libr 1(1):68–88

Aggarwal CC, Wang H (eds) (2010) Managing and mining graph data. Advances in database systems. Springer Science – Business Media, Berlin

Agrawal R, Jagadish HV (1987) Direct algorithms for computing the transitive closure of database relations. In: Proceedings of the international conference on very large data bases (VLDB). Morgan Kaufmann, Los Altos, pp 255–266

Amann B, Scholl M (1992) Gram: a graph data model and query language. In: Proceedings of the European conference on hypertext technology (ECHT). ACM, New York, pp 201–211

Andries M, Gemis M, Paredaens J, Thyssens I, den Bussche JV (1992) Concepts for graph-oriented object manipulation. In: Proceedings of international conference on extending database technology (EDBT). Lecture notes in computer science, vol 580. Springer, Berlin, pp 21–38

Angles R (2012) A comparison of current graph database models. In: 4th international workshop on graph data management: techniques and applications (GDM). ICDE workshop

Angles R, Gutierrez C (2008) Survey of graph database models. ACM Comput Surv 40(1):1–39

Angles R, Barceló P, Ríos G (2013) A practical query language for graph dbs. In: Proceedings of the Alberto Mendelzon international workshop on foundations of data management (AMW)

Angles R, Arenas M, Barceló P, Hogan A, Reutter J, Vrgoĉ D (2017) Foundations of modern query languages for graph databases. ACM Comput Surv 50(5):68

Angles R, Arenas M, Barceló P, Boncz P, Fletcher G, Gutierrez C, Lindaaker T, Paradies M, Plantikow S, Sequeda J, van Rest O, Voigt H (2018) G-core: a core for future graph query languages. In: Proceedings of the international conference on management of data (SIGMOD)

Atre M, Chaoji V, Zaki MJ, Hendler JA (2010) Matrix "bit" loaded: a scalable lightweight join query processor for RDF data. In: Proceedings of the international conference on World Wide Web. ACM, New York, pp 41–50

Barceló Baeza P (2013) Querying graph databases. In: Proceedings of the symposium on principles of database systems (PODS). Invited tutorial. ACM, New York, pp 175–188

Berge C (1973) Graphs and hypergraphs. North-Holland, Amsterdam

Blueprints (2018) https://github.com/tinkerpop/blueprints/wiki

Bornea MA, Dolby J, Kementsietsidis A, Srinivas K, Dantressangle P, Udrea O, Bhattacharjee B (2013) Building an efficient RDF store over a relational database. In: Proceedings of the international conference on management of data (SIGMOD). ACM, New York, pp 121–132

Bray T, Paoli J, Sperberg-McQueen CM (1998) Extensible Markup Language (XML) 1.0, W3C Recommendation. http://www.w3.org/TR/1998/REC-177-19980210

Brijder R, Gillis JJM, Van den Bussche J (2013) The DNA query language DNAQL. In: Proceedings of the international conference on database theory (ICDT). ACM, New York, pp 1–9

Buneman P (1997) Semistructured data. In: Proceedings of the symposium on principles of database systems (PODS). ACM, New York, pp 117–121

Chang CS, Chen ALP (1998) Supporting conceptual and neighborhood queries on the World Wide Web. IEEE Trans Syst Man Cybern 28(2):300–308

Chen PPS (1976) The entity-relationship model - toward a unified view of data. ACM Trans Database Syst 1(1):9–36

Chong EI, Das S, Eadon G, Srinivasan J (2005) An efficient SQL-based RDF querying scheme. In: Proceedings of the international conference on very large data bases. VLDB Endowment, pp 1216–1227

Ciglan M, Averbuch A, Hluchy L (2012) Benchmarking traversal operations over graph databases. In: Proceedings of the international conference on data engineering workshops. IEEE Computer Society, New York, pp 186–189

Codd EF (1970) A relational model of data for large shared data banks. Commun ACM 13(6):377–387

Consens MP, Mendelzon AO (1989) Expressing structural hypertext queries in graphlog. In: Proceedings of the conference on hypertext. ACM, New York, pp 269–292

Consens MP, Mendelzon AO (1990) GraphLog: a visual formalism for real life recursion. In: Proceedings of the symposium on principles of database systems (PODS). ACM, New York, pp 404–416

Consens M, Mendelzon A (1993) Hy+: a hygraph-based query and visualization system. SIGMOD Record 22(2):511–516

Conte D, Foggia P, Sansone C, Vento M (2004) Thirty years of graph matching in pattern recognition. Int J Pattern Recognit Artif Intell 18(3):265–298

Cruz IF, Mendelzon AO, Wood PT (1987) A graphical query language supporting recursion. In: Proceedings of the international conference on management of data (SIGMOD). ACM, New York, pp 323–330

Cruz IF, Mendelzon AO, Wood PT (1989) G+: recursive queries without recursion. In: Proceedings of the international conference on expert database systems (EDS). Addison-Wesley, Reading, pp 645–666

Cudré-Mauroux P, Enchev I, Fundatureanu S, Groth P, Haque A, Harth A, Keppmann F, Miranker D, Sequeda J, Wylot M (2013) NoSQL databases for RDF: an empirical evaluation. In: Proceedings of the international semantic web conference (ISWC). Lecture notes in computer science, vol 8219. Springer, Berlin, pp 310–325

Cypher (2018) http://neo4j.com/developer/cypher-query-language/

Dominguez-Sal D, Martinez-Bazan N, Muntes-Mulero V, Baleta P, Larriba-Pey JL (2010a) A discussion on the design of graph database benchmarks. In: Proceedings of the technology conference on performance evaluation and benchmarking (TPCTC)

Dominguez-Sal D, Urbón-Bayes P, Giménez-Vañó A, Gómez-Villamor S, Martínez-Bazán N, Larriba-Pey JL (2010b) Survey of graph database performance on the HPC scalable graph analysis benchmark. In: Proceedings of the international conference on web-age information management (WAIM). Springer, Berlin, pp 37–48

Dries A, Nijssen S, De Raedt L (2009) A query language for analyzing networks. In: Proceedings of the conference on information and knowledge management (CIKM). ACM, New York, pp 485–494

Elser B, Montresor A (2013) An evaluation study of BigData frameworks for graph processing. In: Proceedings of the international conference on big data. IEEE, New York, pp 60–67

Erling O, Averbuch A, Larriba-Pey J, Chafi H, Gubichev A, Prat A, Pham MD, Boncz P (2015) The LDBC social network benchmark: interactive workload. In: Proceedings of the international conference on management of data. SIGMOD. ACM, New York, pp 619–630

Fan W (2012) Graph pattern matching revised for social network analysis. In: Proceedings of the international conference on database theory (ICDT). ACM, New York, pp 8–21

Faye DC, Cure O, Blin G (2012) A survey of RDF storage approaches. ARIMA J 15:11–35

Gallagher B (2006) Matching structure and semantics: a survey on graph-based pattern matching. In: AAAI fall symposium on capturing and using patterns for evidence detection, pp 45–53

Gemis M, Paredaens J (1993) An object-oriented pattern matching language. In: Proceedings of the international symposium on object technologies for advanced software. Springer, Berlin, pp 339–355

Graves M, Bergeman ER, Lawrence CB (1995) A graph-theoretic data model for genome mapping databases. In: Proceedings of the Hawaii international conference on system sciences (HICSS). IEEE Computer Society, New York, p 32

Gremlin (2018) http://tinkerpop.apache.org/gremlin.html

Guo Y, Biczak M, Varbanescu AL, Iosup A, Martella C, Willke TL (2014) How well do graph-processing platforms perform? an empirical performance evaluation and analysis. In: Proceedings of international parallel and distributed processing symposium. IEEE Computer Society, New York, pp 395–404

Gutiérrez A, Pucheral P, Steffen H, Thévenin JM (1994) Database graph views: a practical model to manage persistent graphs. In: Proceedings of the international conference on very large data bases (VLDB). Morgan Kaufmann, Los Altos, pp 391–402

Güting RH (1994) GraphDB: modeling and querying graphs in databases. In: Proceedings of the international conference on very large data bases (VLDB). Morgan Kaufmann, Los Altos, pp 297–308

Gyssens M, Paredaens J, den Bussche JV, Gucht DV (1990) A graph-oriented object database model. In: Proceedings of the symposium on principles of database systems (PODS). ACM, New York, pp 417–424

Han M, Daudjee K, Ammar K, Özsu MT, Wang X, Jin T (2014) An experimental comparison of pregel-like graph processing systems. Proc VLDB Endow 7(12):1047–1058

Harris S, Seaborne A (2013) SPARQL 1.1 Query Language, W3C Recommendation. https://www.w3.org/TR/sparql11-query/

Harris S, Lamb N, Shadbolt N (2009) 4store: the design and implementation of a clustered RDF store. In: Proceedings of scalable semantic web knowledge base systems (SSWS), pp 94–109

Hayes J, Gutierrez C (2004) Bipartite graphs as intermediate model for RDF. In: Proceedings of the international semantic web conference (ISWC). Lecture notes in computer science, vol 3298. Springer, Berlin, pp 47–61

He H, Singh AK (2008) Graphs-at-a-time: query language and access methods for graph databases. In: Proceedings of the international conference on management of data (SIGMOD). ACM, New York, pp 405–418

Hidders J (2002) Typing graph-manipulation operations. In: Proceedings of the international conference on database theory (ICDT). Springer, Berlin, pp 394–409

Hidders J, Paredaens J (1993) GOAL, a graph-based object and association language. In: Advances in database systems: implementations and applications. CISM. Springer, Wien, pp 247–265

Iosup A, Hegeman T, Ngai WL, Heldens S, Prat-Pérez A, Manhardto T, Chafio H, Capotă M, Sundaram N, Anderson M, Tănase IG, Xia Y, Nai L, Boncz P (2016) LDBC graphalytics: a benchmark for large-scale graph analysis on parallel and distributed platforms. Proc VLDB Endow 9(13):1317–1328

Jouili S, Vansteenberghe V (2013) An empirical comparison of graph databases. In: Proceedings of the international conference on social computing (SocialCom), pp 708–715

Khan A, Elnikety S (2014) Systems for big-graphs. In: Proceedings of the international conference on very large data bases (VLDB)

Kiesel N, Schurr A, Westfechtel B (1996) GRAS: a graph-oriented software engineering database system. In: IPSEN book. Pergamon, New York, pp 397–425

Kim W (1990) Object-oriented databases: definition and research directions. IEEE Trans Knowl Data Eng 2(3):327–341

Klyne G, Carroll J (2004) Resource description framework (RDF) concepts and abstract syntax. https://www.w3.org/TR/2004/REC-rdf-concepts-20040210/

Kotsev V, Minadakis N, Papakonstantinou V, Erling O, Fundulaki I, Kiryakov A (2016) Benchmarking RDF query engines: the LDBC semantic publishing benchmark. In: Proceedings of the workshop on benchmarking linked data, co-located with the international semantic web conference (ISWC)

Kowalik L (2007) Adjacency queries in dynamic sparse graphs. Inform Process Lett 102:191–195

Kunii HS (1987) DBMS with graph data model for knowledge handling. In: Proceedings of the fall joint computer conference on exploring technology: today and tomorrow. IEEE Computer Society Press, Los Alamitos, pp 138–142

Kuper GM, Vardi MY (1984) A new approach to database logic. In: Proceedings of the symposium on principles of database systems (PODS). ACM, New York, pp 86–96

Levene M, Loizou G (1995) A graph-based data model and its ramifications. IEEE Trans Knowl Data Eng 7(5):809–823

Levene M, Poulovassilis A (1990) The hypernode model and its associated query language. In: Proceedings of the Jerusalem conference on information technology. IEEE Computer Society Press, Los Alamitos, pp 520–530

Levene M, Poulovassilis A (1991) An object-oriented data model formalised through hypergraphs. Data Knowl Eng 6(3):205–224

Liu YA, Stoller SD (2006) Querying complex graphs. In: Proceedings of the international symposium on practical aspects of declarative languages. Springer, Berlin, pp 16–30

Low Y, Bickson D, Gonzalez J, Guestrin C, Kyrola A, Hellerstein JM (2012) Distributed GraphLab: a framework for machine learning and data mining in the cloud. Proc VLDB Endow 5(8):716–727

Mainguenaud M (1992) Simatic XT: a data model to deal with multi-scaled networks. Comput Environ Urban Syst 16:281–288

Malewicz G, Austern MH, Bik AJ, Dehnert JC, Horn I, Leiser N, Czajkowski G (2010) Pregel: a system for large-scale graph processing. In: Proceedings of the international conference on management of data (SIGMOD). ACM, New York, pp 135–146

McColl R, Ediger D, Poovey J, Campbell D, Bader DA (2013) A brief study of open source graph databases. http://arxiv.org/abs/1309.2675

McGuinness DL, van Harmelen F (2004) OWL web ontology language overview, W3C recommendation. https://www.w3.org/TR/owl-features/

Mendelzon AO, Wood PT (1995) Finding regular simple paths in graph databases. SIAM J Comput 24(6):1235–1258

Morari A, Castellana V, Villa O, Tumeo A, Weaver J, Haglin D, Choudhury S, Feo J (2014) Scaling semantic graph databases in size and performance. IEEE Micro 34(4):16–26

Neo4j (2018) http://neo4j.com/

Neumann T, Weikum G (2010) The RDF-3X engine for scalable management of RDF data. VLDB J 19(1):91–113

Papadopoulos AN, Manolopoulos Y (2005) Nearest neighbor search - a database perspective. Series in computer science. Springer, Berlin

Papakonstantinou Y, Garcia-Molina H, Widom J (1995) Object exchange across heterogeneous information sources. In: Proceedings of the international conference on data engineering (ICDE). IEEE Computer Society, New York, pp 251–260

Paredaens J, Peelman P, Tanca L (1995) G-Log: a graph-based query language. IEEE Trans Knowl Data Eng 7:436–453

Peckham J, Maryanski FJ (1988) Semantic data models. ACM Comput Surv 20(3):153–189

Poulovassilis A, Levene M (1994) A nested-graph model for the representation and manipulation of complex objects. ACM Trans Inform Syst 12(1):35–68

Prud'hommeaux E, Seaborne A (2008) SPARQL query language for RDF, W3C recommendation. https://www.w3.org/TR/rdf-sparql-query/

Rodriguez MA (2015) The gremlin graph traversal machine and language (invited talk). In: Proceedings of the symposium on database programming languages. ACM, New York, pp 1–10

Rodriguez MA, Neubauer P (2010) Constructions from dots and lines. Bull Am Soc Inf Sci Technol 36(6):35–41

Ronen R, Shmueli O (2009) SoQL: a language for querying and creating data in social networks. In: Proceedings of the international conference on data engineering (ICDE). IEEE Computer Society, New York, pp 1595–1602

Roussopoulos N, Mylopoulos J (1975) Using semantic networks for database management. In: Proceedings of the international conference on very large data bases (VLDB). ACM, New York, pp 144–172

Sakr S, Pardede E (2011) Graph data management: techniques and applications, 1st edn. IGI Global, Hershey

Schmidt M, Hornung T, Küchlin N, Lausen G, Pinkel C (2008) An experimental comparison of RDF data management approaches in a SPARQL benchmark scenario. In: Proceedings of the international semantic web conference (ISWC). Springer, Berlin, pp 82–97

Shipman DW (1981) The functional data model and the data language DAPLEX. ACM Trans Database Syst 6(1):140–173

Stegmaier F, Grobner U, Dolller M, Kosch H, Baese G (2009) Evaluation of current RDF database solutions. In: Proceedings of the international workshop of the multimedia metadata community on semantic multimedia database technologies (SeMuDaTe)

Theodoratos D (2002) Semantic integration and querying of heterogeneous data sources using a hypergraph data model. In: Proceedings of the British national conference on databases (BNCOD). Lecture notes in computer science. Springer, Berlin, pp 166–182

Tian Y, McEachin RC, Santos C, States DJ, Patel JM (2007) Saga: a subgraph matching tool for biological graphs. Bioinformatics 23(2):232–239

Tompa FW (1989) A data model for flexible hypertext database systems. ACM Trans Inform Syst 7(1):85–100

van Rest O, Hong S, Kim J, Meng X, Chafi H (2013) Pgql: a property graph query language. In: Proceedings of the international workshop on graph data management experiences and systems (GRADES)

Vicknair C, Macias M, Zhao Z, Nan X, Chen Y, Wilkins D (2010) A comparison of a graph database and a relational database: a data provenance perspective. In: Proceedings annual southeast regional conference. ACM, New York, pp 1–6

Watters C, Shepherd MA (1990) A transient hypergraph-based model for data access. ACM Trans Inform Syst 8(2):77–102

Weiss C, Karras P, Bernstein A (2008) Hexastore: sextuple indexing for semantic web data management. Proc VLDB Endow 1(1):1008–1019

Wood PT (1990) Factoring augmented regular chain programs. In: Proceedings of the international conference on very large data bases (VLDB). Morgan Kaufmann, Los Altos, pp 255–263

Wood PT (2012) Query languages for graph databases. SIGMOD Record 41(1):50–60

Xin RS, Gonzalez JE, Franklin MJ, Stoica I (2013) GraphX: a resilient distributed graph system on spark. In: Proceedings of international workshop on graph data management experiences and systems (GRADES). ACM, New York, pp 1–6

Yannakakis M (1990) Graph-theoretic methods in database theory. In: Proceedings of the symposium on principles of database systems (PODS). ACM, New York, pp 230–242

Yuan P, Liu P, Wu B, Jin H, Zhang W, Liu L (2013) TripleBit: a fast and compact system for large scale RDF data. Proc VLDB Endow 6(7):517–528

Zeng K, Yang J, Wang H, Shao B, Wang Z (2013) A distributed graph engine for web scale RDF data. Proc VLDB Endow 6(4):265–276

Zhao Y, Yoshigoe K, Xie M, Zhou S, Seker R, Bian J (2014) Evaluation and analysis of distributed graph-parallel processing frameworks. J Cyber Secur Mobil 3(3):289–316

Zhu AD, Ma H, Xiao X, Luo S, Tang Y, Zhou S (2013) Shortest path and distance queries on road networks: towards bridging theory and practice. In: Proceedings of the international conference on management of data (SIGMOD). ACM, New York, pp 857–868

Zou L, Özsu M, Chen L, Shen X, Huang R, Zhao D (2014) gStore: a graph-based SPARQL query engine. VLDB J 23(4):565–590

Chapter 2
Graph Visualization

Peter Eades and Karsten Klein

Abstract Graphs provide a versatile model for data from a large variety of application domains, for example, software engineering, telecommunication, and biology. Understanding the information that is represented by the graph is crucial for scientists and engineers to understand critical issues in these domains. Graph visualization is the process of creating a drawing of a graph so that a human can understand the graph. However, the depth of understanding depends on the quality of the graph representation. Good visualization can facilitate efficient visual analysis of the data to detect patterns and trends. Important aspects of the development of graph drawing methods are the efficiency and accuracy of the algorithms, and the quality of the resulting picture. In this chapter, we discuss the geometric properties of good graph visualizations as node-link diagrams, and describe methods for constructing good layouts of graphs.

2.1 Introduction

Graph visualization is the process of making a drawing of a graph so that a human can understand the graph. This is illustrated as the *graph visualization pipeline* in Fig. 2.1. A *drawing function* D takes a graph G from a graph dataset (a) and produces a graph drawing $D(G)$ (b). A *perception function* P takes the drawing $D(G)$ and produces some knowledge $P(D(G))$ in the human (c). The drawing function D can be executed with pen and paper by a human, but since the advent of computer graphics in the 1970s, there has been increasing interest in executing

P. Eades
University of Sydney, Sydney, NSW, Australia
e-mail: peter.eades@sydney.edu.au

K. Klein (✉)
Monash University, Melbourne, VIC, Australia

University of Konstanz, Konstanz, Germany
e-mail: karsten.klein@uni-konstanz.de

© Springer International Publishing AG, part of Springer Nature 2018
G. Fletcher et al. (eds.), *Graph Data Management*, Data-Centric Systems and Applications, https://doi.org/10.1007/978-3-319-96193-4_2

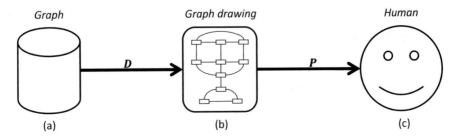

Fig. 2.1 A graph visualization pipeline

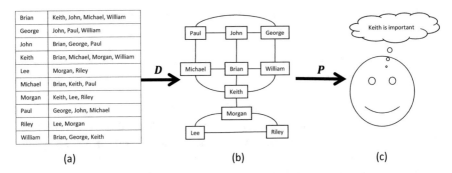

Fig. 2.2 A graph, a drawing of that graph, and some knowledge perceived by the human

this function on a computer; in this chapter, we discuss computer algorithms that implement D. The perception function P is executed by the human's perceptual and cognitive facilities.

As an illustration, consider the social network in Fig. 2.2; it describes a friendship relation between a group of people. It is represented in Fig. 2.2a as a table with the first column listing the people, and the second column listing the friends of each person. For example, the friends of Brian are Keith, John, Michael, and William. The drawing function D produces the picture in Fig. 2.2b; each person is represented by a box with text, and the friendship relation is represented by lines connecting the boxes. The perception function P takes the picture as input and produces some knowledge in the human. This could be low-level knowledge such as "John and George are friends," or higher-level knowledge such as "Keith is important."

Graphs (aka *networks*) are one of the most pervasive models used in technology; social networks are prime examples, but we also find graphs in areas as diverse as biotechnology, in forensics, in software engineering, and epidemiology. For humans to make sense of these graphs, a picture or *graph drawing* is helpful. In this chapter, we introduce the basic methods for creating pictures of graphs that are helpful for humans.

The graph data in Fig. 2.1a is a set of *attributed graphs*. Each such graph consists of a set of vertices (sometimes called "nodes") and a set of binary relationships (often called "edges") between the vertices. The vertices and edges usually have attributes. For example, the vertices in Fig. 2.2 have textual *names*. Edge attributes could include, for example, a number that quantifies the *strength* of a friendship.

The graph drawing in Fig. 2.1b is a "node-link" diagram: it consists of a glyph $D(u)$ for each vertex u of the graph, and a curve segment $D(e)$ connecting the glyphs $D(u)$ and $D(v)$ for each edge $e = (u, v)$ of the graph. Each glyph $D(u)$ has geometric attributes (such as position and size) and graphical attributes (such as color). Similarly, each curve $D(e)$ has geometric attributes (such as its route) and graphical attributes (such as color and linestyle). Note that other kinds of graph drawing are possible; see Sect. 2.5.4 below. However, in this chapter we will concentrate on the node-link metaphor, as it is the most commonly used.

In practice, it is relatively easy to find a good mapping from the vertex and edge attributes to the graphical attributes of glyphs and lines, using well-established rules of graphic design (see, e.g., Tufte 1992). A large variety of graphical notations exist in common application areas. Figures 2.3 and 2.4 show real-world examples of attribute mappings from biology. The representation in Fig. 2.3 uses the *Systems Biology Graphical Notation* (SBGN) standard (Le Novère et al. 2009) and has been produced with *SBGN-ED* (Czauderna et al. 2010), an extension of the *Vanted* framework (Rohn et al. 2012). Figure 2.5 shows a real-world example from biomedicine. The graph represents functional connectivity of brain regions; the color is used to map the correlation of age and functional connectivity in a group of 66 subjects onto the graph. The color coding shows an age-related decrease in

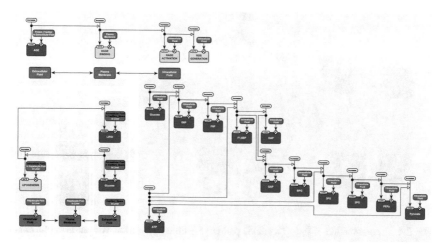

Fig. 2.3 A part of a biological pathway drawn using the SBGN notation. Attributes are mapped to graphical attributes. The network is a part of the visual representation describing the development of diabetic retinopathy, a condition which leads to visual impairment if left untreated

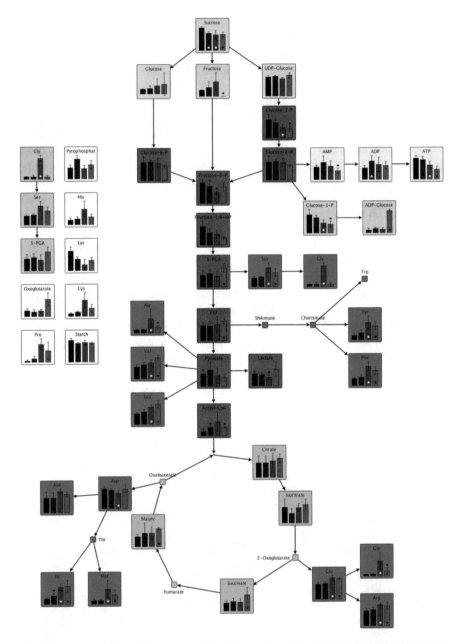

Fig. 2.4 A combination of three metabolic pathways with vertex attributes represented by charts and colors mapped onto the vertex representations. Bar charts represent the amount of a metabolite for four different plant lines

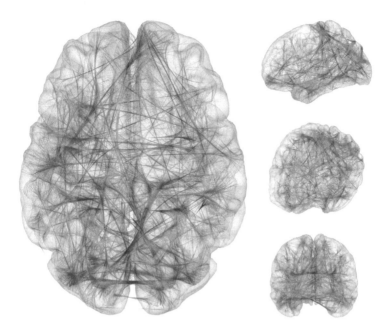

Fig. 2.5 Graph visualization of functional brain region connectivity. Edges connect brain location that have a high activity correlation. Edge bundling is used to reduce clutter and emphasize patterns. Figure as originally published in Böttger et al. (2014)

connection strength in the frontal region, compared to an increase in the central region.

In contrast, it is difficult to find a good layout for a node-link diagram: if we choose the location of each vertex and the route for each edge badly, then the resulting diagram is tangled and hard to read. In Sect. 2.2, we examine the geometric properties of "good" node-link diagrams. Then we describe methods for constructing good layouts of node-link diagrams. In particular, we describe two important approaches: the *topology-shape-metrics* approach (Sect. 2.3), and the *energy-based* approach (Sect. 2.4).

2.2 Readability and Faithfulness

We now consider the properties of "good" drawings of graphs. We concentrate on geometric properties, in particular the location of each vertex and the route for each edge. There are two aspects of the quality of a graph drawing: *readability* and *faithfulness*.

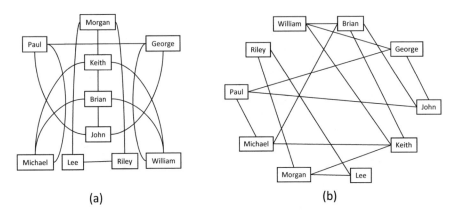

Fig. 2.6 Poor quality drawings of the graph in Fig. 2.2a, (**a**) symmetric and (**b**) circular style

2.2.1 Readability

Readability concerns the quality of the perception function P in Fig. 2.1: how well does the human understand the picture? Two further drawings of the graph in Fig. 2.2a are in Fig. 2.6. Intuitively, these two drawings are less readable than that in Fig. 2.2b.

Geometric properties of readable drawings of a graph are commonly called "aesthetic criteria." Discussions of aesthetic criteria began in the 1970s. For example, Sugiyama (see Sugiyama et al. 1981) and Tamassia et al. (1988) produced structured lists of aesthetic criteria; a sample is given below.

C1 The number of edge crossings is minimized.
C2 The total length of edges is minimized.
C3 The ratio of length to breadth of the drawing is balanced.
C4 The number of edge bends is minimized (using straight lines where possible).
C5 Minimization of the area occupied by the drawing.

All these aesthetic criteria were based on intuition and introspection rather than any scientific evidence. Later, Purchase et al. (1995) began the scientific investigation of aesthetic criteria, based on HCI-style human experiments. She measured the time to complete tasks such as tracing a shortest path in a graph drawing, and errors made in such tasks. These variables were correlated with aesthetic criteria such as those above. Purchase found significant evidence that both time and errors increase with the number of edge crossings and with the number of edge bends, and less significant evidence for other aesthetic criteria. Further experiments (Purchase 2002; Ware et al. 2002; Huang et al. 2014) confirmed, refined, and extended Purchase's original work.

2.2.2   Faithfulness

The work of Nguyen et al. (2013) concerns the quality of the drawing function D in Fig. 2.1. The drawing $D(G)$ of a graph G is *faithful* if it uniquely represents the graph G. In other words, D is faithful if it has an inverse; that is, if the graph G can be recovered uniquely from the drawing $D(G)$. This concept may seem strange at first, because it may seem that all graph drawings are faithful. However, the concept is significant for very large graphs. As a simple example, consider the graph in Fig. 2.7. This drawing uses a technique recently called *edge bundling* (Holten and van Wijk 2009) (originally called *edge concentration* (Newbery 1989)) to cope with the large number of edges. While this drawing may be readable, it is not faithful: it

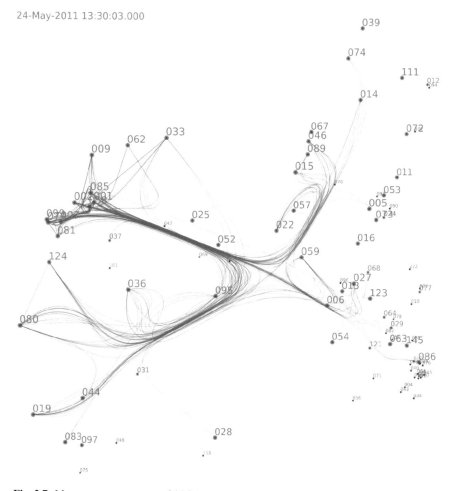

Fig. 2.7 Money movements; an unfaithful graph drawing

does not uniquely represent a graph (because there are many graphs that could have this drawing).

While readability has a long history of investigation, faithfulness has only arisen since the advent of very large data sets, and it is currently not well-understood. One faithfulness criterion that has been proposed (Nguyen et al. 2013) is based on the intuition that in a faithful graph drawing, the distance between u and v in the graph G should be reflected by the geometric distance between the positions $D(u)$ and $D(v)$ of u and v in the drawing. To make this notion more precise, suppose that $\Delta_G(u, v)$ is the distance between u and v in G (e.g., $\Delta_G(u, v)$ could be the length of a graph-theoretic shortest path between u and v). For a drawing function D that maps vertices of a graph $G = (V, E)$ to points in $\mathbb{R}^2$, we define

$$\sigma(D(G)) = \Sigma_{u,v \in V} \left(\Delta_G(u, v) - \Delta_{\mathbb{R}^2}(D(u), D(v)) \right)^2 \qquad (2.1)$$

where $\Delta_{\mathbb{R}^2}$ is a distance function in $\mathbb{R}^2$ (e.g., Euclidean distance). In other words, σ is the sum of squared errors between distances in the graph G and distances in the drawing $D(G)$. In this way, σ measures the faithfulness of the drawings insofar as distances are concerned. In the 1950s, Torgerson (1952) employed a similar criterion when he proposed the *Multidimensional Scaling* method for psychometrics, a projection technique that allows to represent distance information as a two-dimensional (2D) or three-dimensional (3D) picture. Over the following decades, such methods were refined and extended into distance-based graph drawing methods. With the success of the stress minimization approach, these methods have recently gained increasingly interest; see Sect. 2.4.

2.3   The Topology-Shape-Metrics Approach

Motivated by the need to create database diagrams, Batini et al. (1986) introduced a method for drawing graphs. The method has been refined and extended many times, and it is now known as the "topology-shape-metrics" approach. An example of a graph drawing computed with this approach is in Fig. 2.8. The method has three phases:

1. Topology: First, we compute an appropriate topological arrangement of vertices, edges, and faces. In this phase, we aim for a small number of edge crossings.
2. Shape: Next, we compute the general shape of each edge of the drawing. In this phase, we aim for a small number of edge bends.
3. Metrics: Finally, we compute the precise location of each vertex, each edge bend, and each edge crossing. In this phase, we aim for a drawing with high resolution.

Before we describe each of these phases, we outline the concepts of *orthogonal grid drawings* and *planar graphs*; these concepts are needed to understand the approach.

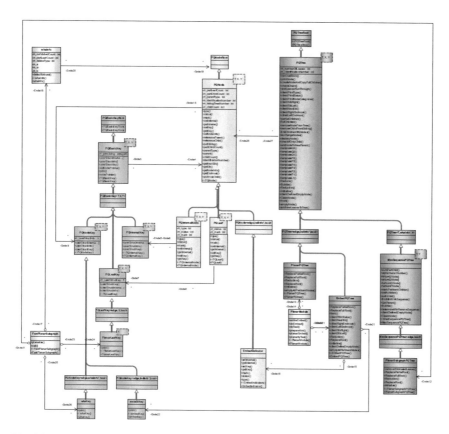

Fig. 2.8 An orthogonal drawing of a UML class diagram, computed using the topology-shape-metrics method in OGDF (as published in Gutwenger et al. (2003), ACM Symposium on Software Visualization 2003)

2.3.1   Orthogonal Grid Drawings

In a *grid drawing* of a graph, each vertex is located at an integer grid point, that is, it has integer coordinates. Examples of grid drawings are in Fig. 2.9. Grid drawings are used to ensure that the drawing has adequate *vertex resolution*, that is, vertices do not lie too close to each other. Suppose that we have a grid drawing in which x_{max}, x_{min}, y_{max}, and y_{min} are the maximum and minimum x and y coordinates of a vertex, respectively. The *area* of the grid drawing is

$$A = (x_{max} - x_{min})(y_{max} - y_{min})$$

and the *aspect ratio* is

$$R = \frac{y_{max} - y_{min}}{x_{max} - x_{min}}.$$

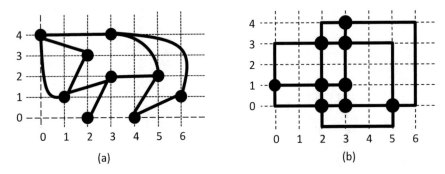

Fig. 2.9 (a) A grid drawing. (b) An orthogonal grid drawing

The drawing in Fig. 2.9a has area $A = 35$ and aspect ratio $R = \frac{5}{7}$. If the drawing is rendered on an $X \times Y$ screen, then the distance between two vertices is at least

$$A^{-0.5} \min \left(X R^{0.5}, Y R^{-0.5} \right);$$

Thus to obtain good vertex resolution, we want a grid drawing in which the area A is as small and the aspect ratio R is close to $\frac{Y}{X}$.

A graph drawing is *orthogonal* if each edge is a polyline consisting of vertical and horizontal line segments. Figure 2.9b shows an orthogonal grid drawing. Note that an orthogonal drawing of a graph with a vertex of degree greater than 4 is necessarily unfaithful; thus, for the moment we assume that every vertex has degree at most 4. In practice, this restriction can be overcome by a variety of methods (such as the *Kandinsky* approach (Fößmeier and Kaufmann 1995)).

Orthogonal drawings are widely used in software design diagrams, such as Fig. 2.8. From the aesthetic criterion C4 in Sect. 2.2, we want an orthogonal graph drawing with few bends.

2.3.2 Planarity and Topology

A drawing of graph is *planar* if it has no edge crossings; a graph G is planar if there is a planar drawing of G. Examples of planar and nonplanar graphs are in Fig. 2.10. The theory of planar graphs has been developed by mathematicians for hundreds of years. For example, Kuratowski (1930) gave an elegant characterization of the class of planar graphs: a graph is planar if and only if it does not contain a subgraph that is a subdivision of the complete graph K_5 on five vertices or the complete bipartite graph $K_{3,3}$ on six vertices (a *subdivision* of a graph is formed by adding vertices along edges). The classical but inelegant linear-time algorithm of Hopcroft and Tarjan (1974) can be used to test whether a graph is planar; simpler linear-time

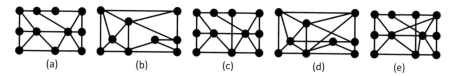

Fig. 2.10 The graph drawings (**a**) and (**b**) are planar, and the graph drawings (**c**), (**d**), and (**e**) are nonplanar. However, the graph in (**c**) is planar because there is a planar drawing of this graph. The graphs in (**d**) and (**e**) are nonplanar

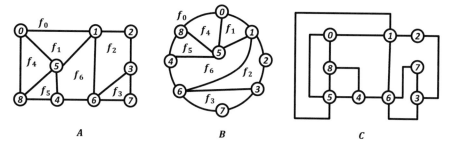

Fig. 2.11 The graph drawings (**a**) *and* (**b**) *have the same topology. The graph drawing* (**c**) *displays the same graph, but the topology is different*

algorithms have been developed more recently (Boyer and Myrvold 2004b; Shih and Hsu 1999).

A planar graph drawing divides the plane into regions called *faces*. The drawing *A* in Fig. 2.11 has seven faces $f_0, f_1, \ldots, f_6$ (note that f_0 is the *outside face* of the drawing). Two drawings of a graph *G* are *topologically equivalent* if there is a continuous deformation of the plane that maps one to the other; an equivalence class under topological equivalence is a *topological embedding* of *G*. To illustrate this, consider the three planar drawings *A*, *B*, and *C* of a graph in Fig. 2.11. Here *A* and *B* have the same topological embedding: it is possible to deform the plane so that *A* transforms to *B*. It can be seen that *C* has a different topological embedding, because while *A* and *B* both have a face with eight vertices, *C* has no such face.

It is well-known that two planar drawings of the same graph are topologically equivalent if and only if the clockwise circular order of edges around each vertex is the same. One can check this property for the examples in Fig. 2.11. This property is a combinatorial characterization of a topological embedding, and can be used to construct data structures that implement operations on topological embeddings efficiently (see, e.g., Chrobak and Eppstein 1991). Variations of the Hopcroft–Tarjan algorithm (see, e.g., Mehlhorn and Mutzel 1996) can be used to construct a topological embedding (as a clockwise circular ordering of edges around each vertex) of a planar graph in linear time.

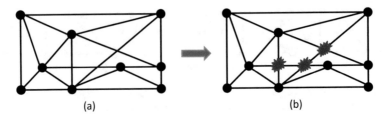

Fig. 2.12 Adding dummy vertices to the nonplanar graph drawing (**a**) gives the planar drawing (**b**)

2.3.3   Computing the Topology, Using Planarization

A nonplanar graph drawing can be converted into a planar graph drawing simply by adding new "dummy" vertices at each crossing point, as illustrated in Fig. 2.12. This simple process of adding dummy vertices gives the intuition for the "topology" phase of the topology-shape-metrics method. However, graph visualization begins with a graph, not with a graph drawing, and converting to a planar graph is not so straightforward. Further, from the aesthetic criterion C1 in Sect. 2.2, we want to ensure that the number of crossings is small.

The "topology" phase, sometimes called a *planarization* process, takes a nonplanar graph $G = (V, E)$ as input, and produces a planar topological embedding $G''' = (V''', E''')$ as output. The first step is to find planar subgraph $G' = (V, E')$ ($E' \subset E$) where $|E'|$ is as large as possible. This is a nontrivial problem; in fact, finding a maximum planar subgraph of a given graph is NP-hard (Garey and Johnson 1979). However, a number of heuristic methods are available (Jünger et al. 1998; Jünger and Mutzel 1994, 1996). The next step is to find a topological embedding G'' of the planar graph G'. This step is relatively easy, and can be accomplished in linear time using a variation of the Hopcroft–Tarjan algorithm, or perhaps one of the simpler algorithms developed more recently (e.g., see Boyer and Myrvold 2004a; Shih and Hsu 1999). The third step is to insert the edges of $E - E'$. The aim in this step is to minimize the number of crossings; although this is NP-hard, it is common to use the simple strategy of inserting one edge at a time, locally minimizing crossings at each insertion. The local minimization can be done by using a shortest path algorithm on the graph of faces of G''. This gives our planar topological embedding $G''' = (V''', E''')$.

We can illustrate the planarization process with an example. We begin with a graph G, represented as a table in Fig. 2.13a; note that this is a combinatorial graph, with no topology or geometry. A (bad) drawing of this graph is in Fig. 2.13b. Further, G is a nonplanar graph by Kuratowski's Theorem, because there is a subgraph (shown in Fig. 2.13c) that is a subdivision of the complete graph K_5 on five vertices. Next, we identify a large planar subgraph G' of G, using one of the heuristic methods available. In this case, we can delete the edges $(2, 6)$ and $(6, 7)$ to give a planar subgraph. Using a variation of the Hopcroft–Tarjan algorithm, we can find

A graph **G**

Nodes	Adjacent nodes
0	2,5,6,8
1	3,4,7
2	0,3,6,8
3	1,2,4,7
4	1,3,5
5	0,4,6,8
6	0,2,5,7
7	1,3,6,8
8	0,2,5,7

(a)

A drawing of **G**

(b)

A subdivision of K_5 in **G**

(c)

Fig. 2.13 A nonplanar graph

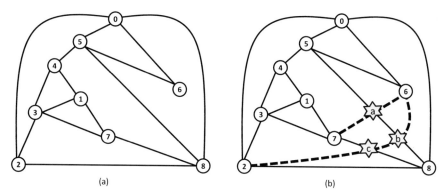

(a)   (b)

Fig. 2.14 (**a**) A topological embedding G'' of G'. (**b**) The edges $(2, 6)$ and $(6, 7)$ are re-inserted, with dummy vertices a, b, and c, to form a topological embedding G'''

a topological embedding G'' of G'. Such an embedding is illustrated in Fig. 2.14a. Finally, we reinsert the edges $(2, 6)$ and $(6, 7)$, and place dummy vertices a, b, and c at the crossing points, to give a topological embedding G''', as in Fig. 2.14b.

2.3.4 Computing the Shape

The output of the topology phase is a topological embedding, which we shall now denote as G. Some of the vertices of G are dummy vertices, representing crossings between edges in the original graph. The final drawing output from topology-shape-metrics method is an orthogonal drawing, in that each edge is a sequence of horizontal and vertical line segments. The shape phase chooses "shape" of each edge, in the following sense. Suppose that the edge (u, v) is directed from u to v, and it consists of a sequence $(u_0, u_1), (u_1, u_2), \ldots, (u_{k-1}, u_k)$ of k segments, where $u_0 = u$ and $u_k = v$. Each line segment (u_i, u_{i+1}) has a compass direction: either *north*, *south*, *east*, or *west*. The sequence $(d_0, d_1, \ldots, d_{k-1})$, where d_i is the

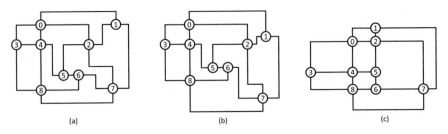

Fig. 2.15 Three orthogonal drawings with the same topological embedding. The drawings in (**a**) and (**b**) have the same shape, and each has 19 edge bends; the drawing in (**c**) has a different shape and has eight edge bends

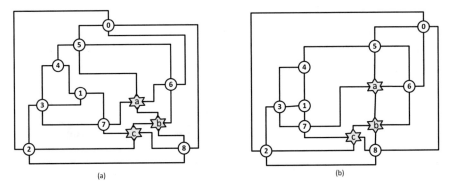

Fig. 2.16 Two orthogonal drawings with the topological embedding in Fig. 2.14. The shape (**a**) has more bends than (**b**)

compass direction of (u_i, u_{i+1}), is the *shape* of the edge (u, v). As examples, the edge $(0, 1)$ in Fig. 2.15a has shape *(north, east, south)*, and the edge $(1, 2)$ has shape *(west, south, west)*.

Two orthogonal drawings A and B have the *same shape* if each edge in A has the same shape as the corresponding edge in B. The drawing in Fig. 2.15a has the same shape as that in Fig. 2.15b. However, Fig. 2.15c has a different shape.

Note that drawings in Fig. 2.15a, b have 19 edge bends each, but Fig. 2.15c has only eight. The aim of the shape phase is to choose a shape with few bends.

A surprising result of Tamassia (1987) gives a polynomial-time algorithm for choosing a shape with a minimum total number of edge bends. Tamassia's algorithm is based on a reduction to the maximum flow problem; the best implementation (Garg and Tamassia 1996) runs in time $O(|V|^{1.75})$. A simpler algorithm of Tamassia and Tollis (1986), based on so-called visibility graphs, runs in linear time and results in a drawing with at most four bends per edge (but not necessarily giving a minimum total number of bends).

A naïve routing of orthogonal edges for the topological embedding illustrated in Fig. 2.14 is given in Fig. 2.16a. A better shape for this embedding is in Fig. 2.16b.

2.3.5 Computing the Metrics

The shape phase described above determines the sequence of horizontal and vertical line segments that make up each edge. The metrics phase chooses integer coordinates for each vertex, each bend, and each crossing point. Each of these points is located at an integer grid point, and thus we have an orthogonal grid drawing. The main aim of the metrics phase is to give a drawing of small area; the phase is sometimes called *compaction*. An example is in Fig. 2.17. The problem of constructing a layout with small area has a long history in the literature of VLSI layout, and methods can be borrowed.

As the final step of the metrics phase, the graph is rendered without rendering the dummy vertices. The final drawing for the graph in Fig. 2.13 is in Fig. 2.18.

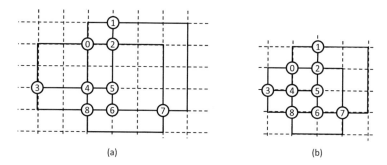

(a)   (b)

Fig. 2.17 Two orthogonal grid drawings with the same shape. The drawing (**a**) has a larger area than (**b**)

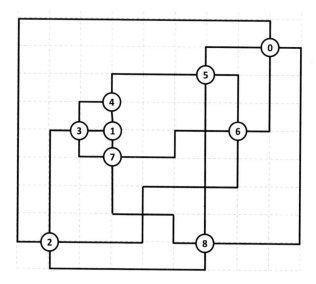

Fig. 2.18 Final drawing for the graph in Fig. 2.13

2.3.6 Remarks and Open Problems for the Topology-Shape-Metrics Approach

The topology-shape-metrics approach has been improved, refined, and extended many times since its inception. Figures 2.3 and 2.8 are examples of the output of such algorithms. The methods work well for small-to-medium-sized orthogonal graph drawings, but are less successful on large graphs.

Nevertheless, a number of open problems remain:

1. *Clustered planarity.* A common method for dealing with very large data sets is to cluster the data, then treat each cluster as a data item. This method is also used in graph drawing: the vertices of a very large graph can be clustered to form "super-vertices"; these super-vertices can be clustered to form "super-super-vertices," and so on, in a hierarchical fashion. More formally, a *clustered graph* $C = (G, T)$ consists of a graph G and a rooted tree T such that the leaves of T are the vertices of G. The tree T forms a cluster hierarchy on the graph. A drawing of a clustered graph $C = (G, T)$ consists of a drawing of the graph G and a region $r(t)$ of the plane for each vertex T of the tree T, such that

    (a) If t_1 is a child of t_0 in T, then $r(t_1) \subset r(t_0)$.
    (b) If t_1 is not a descendent of t_0 and t_0 is not a descendent of t_1 in T, then $r(t_1) \cap r(t_0) = \emptyset$.
    (c) If u is a vertex of G (and thus a leaf of T) then the location of u in the drawing of G is inside $r(u)$.
    (d) If (u, v) is an edge of G and the curve representing (u, v) intersects $r(t)$ for some vertex t of T, then either u or v is a descendent of t.
    (e) If (u, v) is an edge of G and both u and v are descendants of t, then the curve representing (u, v) is inside $r(t)$.

    Further, the drawing of C is *clustered-planar* if the drawing of G is planar. An example of a drawing of a clustered graph $C = (G, T)$ is in Fig. 2.19a; note that this drawing is not clustered-planar. The tree T is illustrated in Fig. 2.19b. Note that although the underlying graph G is planar (see Fig. 2.20), there is no

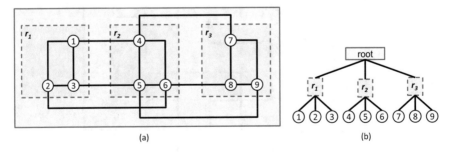

(a)   (b)

Fig. 2.19 (a) A drawing of a clustered graph $C = (G, T)$; (b) the tree T

Fig. 2.20 The underlying graph G for the clustered graph in Fig. 2.19

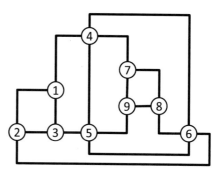

Fig. 2.21 The Mutzel experiment

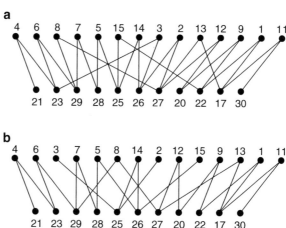

clustered-planar drawing of this clustered graph. (Observe that in any planar embedding of G, the three-cycle $(4, 5, 6)$ must have either at least one of the three-cycles $(1, 2, 3)$ or $(7, 8, 9)$ inside; thus, the cluster region r_2 would have to contain either cluster region r_1 or r_2, contradicting the above rules for clustered drawings.) A clustered graph is *clustered-planar* if it has a clustered-planar drawing. Clustered planarity is a significant problem: shape/metrics steps for clustered graphs are well-established, but despite much investigation (Eades et al. 1999; Feng et al. 1995; Jelínková et al. 2009; Dahlhaus 1998; Cortese and Di Battista 2005; Gutwenger et al. 2002; Cortese and Di Battista 2005; Chimani and Klein 2013; Chimani et al. 2014), the planarization step is still unsolved.

2. *Different ways to count edge crossings.* In the mid-1990s, Mutzel performed an informal experiment during a lecture at a conference. She showed the audience two drawings of the same graph, shown in Fig. 2.21a, b. The audience overwhelmingly preferred (a), despite the fact that it has significantly more edge crossings than (b). In fact, most of the audience mistakenly assumed that (b) had fewer edges than (a). Mutzel's experiment challenged the conventional wisdom that simply counting the number of edge crossings gives a good metric for

Fig. 2.22 A 1-planar drawing. Note that all edge crossings are at right angles

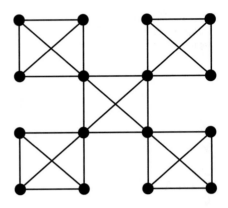

the quality of a graph visualization, and led to a number of new directions for research:

- A number of researchers (Auer et al. 2015; Brandenburg 2014; Giacomo et al. 2014; Sultana et al. 2014; Eades and Liotta 2013; Eades et al. 2013) have begun to investigate *k-planar drawings* where the number of crossings *on each edge* is at most *k*. For example, the drawing in Fig. 2.22 is 1-planar. Most of this research concentrates on mathematical properties of *k*-planar drawings with $k = 1$ or $k = 2$; a good practical algorithm for finding a *k*-planar drawing with minimum *k* remains unknown.

- Huang et al. (2014) showed that edge crossings are tolerable if the crossing angle is large. For example, all crossings in the drawing in Fig. 2.22 are at right angles. For *orthogonal* drawings, all crossing angles are right angles, and perhaps the number of crossings is not significant at all! (See Biedl et al. (1997) for an orthogonal drawing algorithm that ignores edge crossings.) Current research has been mostly mathematical (see Argyriou et al. 2013; Arikushi et al. 2012; Didimo et al. 2009), and the investigation of good practical methods for drawing with large crossing angles is just beginning.

2.4 Energy-Based Approaches and Stress Minimization

The most popular approach to create a layout for undirected graphs is based on so-called *energy-based layout methods*. This popularity is due to the intuitive underlying model of the basic versions, and the fact that these methods can be reasonably easy to implement. In addition, the resulting layouts are often aesthetically pleasing, drawings are described to have a more "organic" or natural appearance than drawings from other methods, and that they show symmetries well. Edges are normally represented as straight lines, which makes bend minimization unnecessary. Figure 2.23 shows an example drawing created with an energy-based method in comparison to an orthogonal drawing.

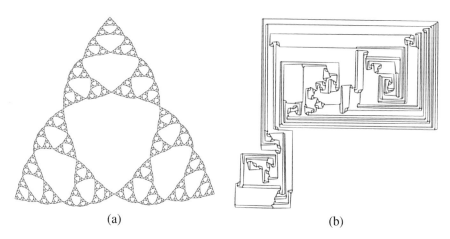

(a)   (b)

Fig. 2.23 Drawing of a Sierpinski triangle, a fractal defined as a recursively subdivided triangle. (a) Drawing created by an energy-based method (b) Drawing created with the topology-shape-metrics approach described in Sect. 2.3

The underlying concept of energy-based methods is to model the graph as a system of objects that contribute to the overall "energy" of the system, and energy-based methods then try to minimize the energy in the system. A basic assumption for the success of such an approach is that a low energy state of the system corresponds to a good drawing of the graph. In order to achieve such an optimum, an energy function is minimized. Energy-based methods thus consist of two main components: a model of objects and their interactions (a virtual physical model), and an algorithm to compute a configuration with low energy (an energy minimization method). There are various models and algorithms under this approach, and the flexibility in the definition of the energy model and energy function implies a wide range of both optimization methods.

Energy-based drawing methods have a long history. Tutte (1960, 1963) used such an approach in one of the earliest graph drawing methods, based on barycentric representations that are obtained by solving a system of linear equations. Tutte proposed the *barycenter algorithm* to draw a triconnected planar graph $G = (V, E)$, and showed that the result is a planar drawing where every face is convex. The algorithm proceeds by first selecting a subset A of the vertices of the graph G to constitute the outer face of the topological embedding of G. The vertices of A are placed on the apices of a convex polygon, and are fixed. Each remaining vertex is placed so as to minimize an energy function that simulates a system of elastic bands. In fact, minimum energy is obtained when each vertex in $V - A$ is at the barycenter of its graph-theoretic neighbors. This setting can be modeled by a nondegenerate system of linear equations, where the position of each vertex is determined as a convex combination of its neighbors' positions. Such a system has a unique solution that can be computed in polynomial time. Barycenter drawings can be very beautiful. However, many barycenter drawings have poor *vertex resolution*,

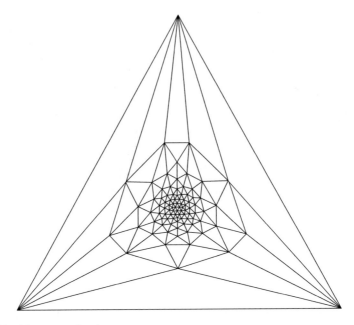

Fig. 2.24 A barycenter drawing

in the sense that vertices can be placed very close to each other. See Fig. 2.24 for an example.

Implementations of energy-based layout methods can be found in a large number of software tools and web services, and most drawings published in both a scientific and nonscientific context are computed using such methods. Some of the more sophisticated energy-based methods allow us to compute layouts for graphs with several hundreds of thousands of vertices in seconds on a standard desktop machine.

A classical, and still frequently used, example for energy-based methods are the so-called *force-directed models*, where the graph objects are modeled as physical objects that mutually exert forces on each other. In the most simple model, unconnected vertices repel each other, and vertices linked by edges attract each other. Force-directed methods have also been applied early for printed circuit board design, where a system of elastic leads and repulsive forces was described for the construction of circuit board drawings. For example, the spring embedder model (Eades 1984) models vertices as electrically charged steel rings and edges as springs, such that the electrical repulsion between vertices and the mechanical forces exerted by the springs in a given layout define the energy of the system (see Fig. 2.25). A minimization of the overall system energy is associated with a layout that optimizes the Euclidean distances between the vertices with respect to a given ideal distance.

The minimization is done in an iterative fashion, moving toward a local energy minimum. First, the vertices are placed in an initial layout, and then in each

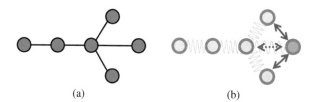

(a)   (b)

Fig. 2.25 Applying the spring embedder model (**b**) to the graph in (**a**). Vertices are modeled by steel rings, edges by springs. Springs exert a force when their length deviates from the natural spring length, which is a parameter for the model, and vertices repel each other. In (**b**), repulsion forces from the darker shaded steel ring are represented by arrows. The force decreases with the square of the distance between vertex pairs

iteration a displacement is computed for each vertex based on the forces exerted on it by vertex repulsion and edge attraction. At the end of the iteration, the positions of all vertices are updated, and a new iteration is started unless the overall displacement falls under a certain threshold. The spring between vertices u and v has an ideal length ℓ, and in a given layout this spring has a current length $\Delta_{\mathbb{R}^2}(u, v)$ (the Euclidean distance between u and v). A variation of Hooke's law is applied to compute the force exerted by a spring, based on the relation between ℓ and $\Delta_{\mathbb{R}^2}(u, v)$: if $\Delta_{\mathbb{R}^2}(u, v)$ is larger than ℓ, then the vertices are attracted to each other, and if it is smaller then they are repelled.

The iterations can be continued until the total force on each vertex converges to zero. In practice, the number of iterations may be limited to a bound K that depends on the size of the graph; then the runtime is $O(K|V|^2))$.

While the first energy-based methods and models were intuitive and rather simple, and the corresponding methods were widely successful in practice, they also exhibit certain disadvantages. First of all, they are rather slow and do not scale well to graphs with more than a few hundred vertices. They are thus not well-suited to cope with the much larger graphs that are visualized today, such as protein–protein interactions in biology or social interactions in social network analysis, with thousands to millions of vertices and edges. Secondly, they rely on an initial drawing and tend to get stuck in a local energy minimum during optimization; see Fig. 2.26 for examples.

Recent approaches, discussed in Sect. 2.4.1, are more complex and make use of more advanced mathematical methods for the optimization. This development allowed large improvements both in the drawing quality and in the computational efficiency.

2.4.1 Scaling to Large Graphs

Three major directions can be identified that in recent years have led to large improvements both regarding the layout quality and the runtime performance. The

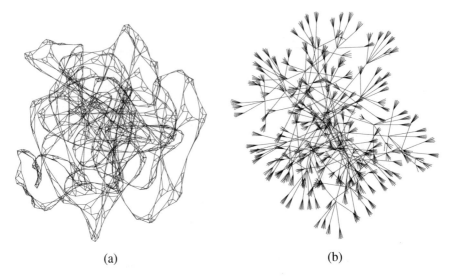

(a)   (b)

Fig. 2.26 Typical unfolding and convergence problems for iterative force-directed algorithms: A Sierpinski triangle with 1095 vertices (**a**) and a tree with 1555 vertices (**b**). Even though both graphs are planar and sparse, no planar drawing with good structure representation was computed

first one is the emergence of so-called *multilevel methods*. The second one is the improvement in optimization process, in particular *fast approximations* for the energy computation. The third one is the identification of *better energy functions*.

2.4.1.1 Multilevel Methods

The multilevel paradigm is a generic approach to handle large datasets by reducing the complexity over a number of hierarchically ordered levels. It is well-suited for graph algorithms and can be used to improve energy-based layout methods, regarding both the layout quality and computation time. The first use of the multilevel approach is commonly credited to Barnard and Simon (1994), where it was used to speed up the recursive spectral bisection algorithm. In the context of graph partitioning, Karypis and Kumar (1995) showed that the quality of the multilevel approach can also be theoretically analyzed and verified.

Multilevel layout methods consist of three components, *coarsening*, *single-level layout*, and *placement*. The main idea is to construct a sequence of increasingly smaller graph representations ("coarsening levels") that approximately conserve the global structure of the input graph G, and to then compute a sequence of approximate solutions, starting with the smallest representation. Intermediate results can be used on the subsequent level to speed up the computation and to achieve a certain quality (Fig. 2.27).

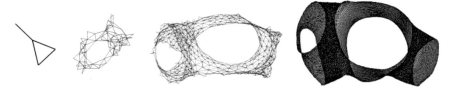

Fig. 2.27 Several levels during multilevel layout computation for a graph with 11,143 vertices and 32,818 edges. The leftmost drawing shows the coarsest level, the rightmost the final drawing. The graph is part of the 10th DIMACS implementation challenge, and available from the UFL Sparse Matrix Collection (University of Florida 2015)

The graph representations are typically created by a series of graph contractions, where a set of vertices on one level is collapsed to a single representative on the next, smaller level. A contraction operation can, for example, simply be an edge contraction, that is, a set of two adjacent vertices is collapsed, and the coarsening for one level then includes contractions of all edges from a maximum independent edge set. These contractions are repeated until the graph size reduces to a given threshold. For graph drawing purposes, usually a threshold of 10–25 vertices is chosen, where force-directed methods can achieve a high-quality drawing quickly. For the resulting levels of the *coarsening phase* now *single-level layouts* are calculated. After computing a layout for the coarsest level from scratch, for each of the intermediate levels a force-directed layout method is applied. As each vertex v from the coarser level l_i represents a set of vertices s on the current, finer level l_{i-1}, the layout for l_i can be used to create an initial drawing for l_{i-1} that is then iteratively improved. Initial positions for vertices can be derived from l_i during this placement phase by simple strategies, such as placing vertices at the barycenter of their neighbors. Experiments indicate that the influence of different placement strategies is marginal (Bartel et al. 2010).

Although several layouts have to be computed, including one for the original graph, the reuse of intermediate drawings leads to less required work and much better convergence on each level than for single-level methods.

Walshaw (2003), Harel and Koren (2002), and Gajer et al. (2004), introduced the multilevel paradigm to graph drawing, after a closely related concept, the *multiscale method*, was proposed by Hadany and Harel (2001). Another related method, the FADE paradigm (Quigley and Eades 2001), used a geometric clustering of the vertex locations for coarsening.

Multilevel approaches can help to overcome local minima and slow convergence problems by improving the unfolding process due to a good coarsening and subsequent placement. However, while multilevel methods can cope even with very large graphs, it may still happen that the resulting layout represents a local minimum far from the optimum. Hachul and Jünger (2007) presented an experimental study of layout algorithms for large graphs, including energy-based multilevel approaches. Bartel et al. (2010) presented an experimental comparison of multilevel layout methods within a modular multilevel framework.

2.4.1.2 Fast Approximations

Another important concept for the practical improvement of energy-based methods is the approximation of the forces to speed up the force calculation. Typically, the repulsive forces are computed approximately, whereas the attraction forces are computed exactly. This means that all edges are taken into account, but individual forces are not calculated for every pair of vertices, as this would mean a runtime of $\Omega(|V|^2)$. One of the first such approximations was the grid-based variant of the Fruchterman–Reingold algorithm (Fruchterman and Reingold 1991), which divides the display space into a grid of squares and for each vertex restricts repulsive forces to vertices within nearby squares. More sophisticated versions (Hachul and Jünger 2004; Quigley and Eades 2001) involve the application of space decomposition structures such as *quadtrees* (see Fig. 2.28) for geometric clustering, as well as efficient approximation schemes such as the *multipole method* (e.g., see Yunis et al. 2012).

Hachul and Jünger combine a multilevel scheme with the multipole approximation, leading to a very fast layout algorithm. With an asymptotic runtime of $O(|V| \log |V| + |E|)$, in practice the algorithm is capable of creating high-quality drawings of graphs up to 100,000 vertices in around a minute.

Spectral graph drawing methods are fast algebraic methods that compute a layout based on so-called *eigenvectors*, sets of vectors associated with matrices defined by the adjacency relations in a graph. Spectral drawing methods were introduced by Hall (1970) in the 1970s. Later developments include the algebraic *multigrid method* ACE (Koren et al. 2002) and the high-dimensional embedding approach HDE (Harel and Koren 2004). These methods show that algebraic methods are very fast and can create reasonable layouts for a variety of graph classes. However, these methods tend to hide details of the graph and are prone to degenerative effects for some graph classes, such as where large subgraphs are projected on a small strip of the drawing area. Hachul and Jünger's experimental study of large graph layout

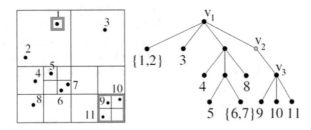

Fig. 2.28 Use of a quadtree for space partitioning, as shown in Hachul and Jünger (2004). First the drawing space is recursively partitioned into four squares, until each square only contains a few vertices (left). The resulting hierarchy can be efficiently represented by a quadtree structure (right), which in turn can be used to allow an efficient force approximation. Forces that act on a vertex are only calculated directly for close-by vertices, whereas the force contribution of a group of vertices that is in a faraway partition is only approximated, replaced by a group force. In the left drawing, the impact of vertices 9, 10, and 11 on vertex 1 is combined in an approximated group force

methods, which also includes algebraic approaches, consequently shows that, while being very fast, algebraic methods often fail to compute reasonable drawings.

2.4.1.3 Better Energy Functions

Distance-based drawing methods constitute an alternative perspective to the graph layout problem. They aim to create a faithful projection from a high-dimensional space to 2D or 3D, with the dimensions simply arising from some notion of dissimilarity, or distance, of each vertex to all other vertices. To this end, distance-based methods usually minimize the *stress* in the drawing, which measures the deviation of the vertex pair distances in the drawing to their dissimilarity. The objective function for stress minimization is

$$\Sigma_{u,v \in V} w_{uv} \left(\Delta_*(u, v) - \Delta_{\mathbb{R}^2}(D(u), D(v)) \right)^2 \qquad (2.2)$$

which sums up the stress for all pairs of vertices u,v located at positions $D(u)$ and $D(v)$, respectively, where $d_{uv} \Delta_*(u, v)$ is the *desired* distance between u and v. Note the similarity between Eqs. (2.1) and (2.2); we can regard stress minimization methods as faithfulness maximization methods. The value w_{uv} is a normalization constant that is often set to $\frac{1}{\Delta_G(u,v)^2}$, where $\Delta_G(u, v)$ is the length of the shortest path between u and v. This emphasizes the influence of deviations from the desired distance for pairs of vertices that have a short graph theoretic distance, and the influence is dampened with increasing distances between pairs of vertices.

The energy-based approach by Kamada and Kawai (1989) uses the shortest graph-theoretic paths as ideal pairwise distance values and subsequently tries to obtain a drawing that minimizes the overall difference between ideal and current distances in an iterative process. Kamada and Kawai propose to use a two-dimensional Newton–Raphson method to solve the resulting system of nonlinear equations in a process that moves one vertex at a time to achieve a local energy minimum. As the cost function involves the all-pairs shortest-path values, the complexity is at least $O(|V|^2 log|V| + |V||E|)$ or $|V|^3$ for weighted graphs, depending on the algorithm used, and $O(|V|^2)$ for the unweighted variant. Following the Kamada–Kawai model, *stress majorization* was introduced as an alternative and improved solution method (Gansner et al. 2004). Several improvements were proposed to make such methods more scalable, for example, by approximation of the distances (Khoury et al. 2012), adding an entropy model (Gansner et al. 2013b), or to respect nonuniform edge lengths (Gansner et al. 2013a).

Each of these improvements led to methods that clearly outperform their predecessors in runtime and layout quality. The combination of the multilevel approach and the multipole method for force approximation in the Fast Multiple Multilevel Method FMMM (Hachul and Jünger 2004), or the use of the maxent-stress model (Gansner et al. 2013c), efficiently computing drawings that clearly depict the structure of many graphs up to a size of several thousand vertices and edges.

2.4.2 Constraint-Based Layout Using Stress

Even though not suited for drawings of large graphs, *constraint-based* drawing methods constitute the most flexible approach to draw graphs in practice. Constraint-based methods take a declarative approach to graph drawing (Lin and Eades 1994); that is, instead of giving an algorithm that describes how to compute a drawing, requirements for the drawing are defined, for example, by geometric constraints. They often resort to generic solution techniques, such as integer linear programming (ILP) or constraint programming (CP) methods, to solve the resulting system of constraints, and are thus often significantly slower than other approaches.

The big advantage of constraint-based methods is that they are able to create high-quality drawings for small-to-medium graphs while taking into account user-defined constraints, such as requirements from a drawing convention, grouping, node sizes and orientation, or personal preferences (see Fig. 2.29).

As a result, such methods are getting increasingly popular in application areas where highly constrained drawings are required; for example, where representation of structural and semantical information beyond the basic structure is needed. Applications include the drawing of technical and flow diagrams to depict hardware systems or biological processes (Schreiber et al. 2009; Rüegg et al. 2014). In the last years great progress has been made in such methods to allow interactive graph layout, in commonly used environments such as web browsers (Monash University 2015).

While constraint-based methods are well-suited for relatively easy extension of a drawing approach by additional drawing constraints, this comes at the cost of decreased computational efficiency of the resulting approach. While there have been approaches to speed up the optimization for specific constraint types (Dwyer 2009), the runtime performance is still an impediment for a more widespread use of constraint-based methods. In addition, to guarantee a certain quality of the computed drawings a good compromise has to be found to prioritize more important soft constraints over less important ones, and a conflict-solving strategy has to be employed.

For further reading, see the surveys in Kobourov (2013), Hu and Shi (2015).

2.4.3 Remarks and Open Problems for Energy-Based Methods

Energy-based methods, in one form or another, are well-established tools for graph visualization. Nevertheless, many open problems remain:

1. *Animation.* An important advantage of energy-based methods, based on the iterative nature of the numerical methods to compute the layout, is that they allow a smooth *animation* of the change from one drawing to another (since the energy function is smooth). They provide one kind of solution to the so-called mental map problem (Misue et al. 1995). Although existing graph drawing tools often

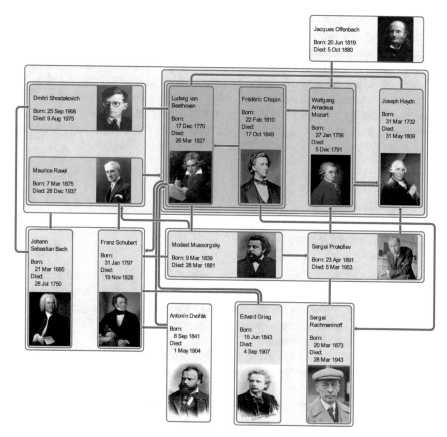

Fig. 2.29 Links between composers, ultra-compact grid layout created using a constraint-based method (Yoghourdjian et al. 2016). For each composer, biographical information and a portrait is shown, the layout system automatically chooses the orientation of the nodes to minimize area usage while optimizing further quality metrics like stress. A hierarchical grouping is computed that groups nodes with similar connections into the same group (hierarchy level represented by color saturation). The number of required edges is thereby reduced, as edges can attach to groups and thus are shared by all members of the group (As published in Yoghourdjian et al. (2016), ©IEEE 2015)

use this kind of animation, it has received little attention from researchers. A thorough investigation of energy-based animation methods would be useful.

2. *Why are some graphs hard to draw?* Energy-based methods are successful on many graphs, but unsuccessful on many others. Intuitively, some graph-theoretic properties are behind success or failure; for example:

- *Dense* graphs (i.e., graphs with many edges) often lead to a "hairball" drawing that makes it hard to perceive the graph structure.
- *Low diameter* graphs seem to become cluttered with distance-based methods.

It would be useful to justify this intuition with mathematical theorems.

3. *Edge crossings and energy-based methods.* It is commonly claimed (without justification) that energy-based methods reduce edge crossings. However, in some cases it seems impossible to avoid crossings with energy-based methods, even on edges of planar subgraphs (Angelini et al. 2013). The relationship between low-energy drawings and edge crossings needs investigation, both mathematically and empirically. (Note that recent results seem to indicate, however, that crossings might not be the dominating factor regarding readability of large graphs; see (Eades et al. 2015; Kobourov et al. 2014)).

2.5   Further Topics

2.5.1   *Directed Graphs*

The methods described in Sects. 2.3 and 2.4 can be used to draw directed graphs (as in Fig. 2.8), but these methods ignore the directions on the edges. It can be useful to have the arrows representing directed edges laid out so that the general "flow" is from one side of the screen to another. For example, in Fig. 2.30, the "flow" is mostly from the top to the bottom.

Sugiyama et al. (1981) described a method for drawing directed graphs. The vertex set is divided into in "layers," and each layer is drawn on a horizontal line; in Fig. 2.30, there are four layers. The layers are chosen so that there are not too many layers, the number of vertices in each layer is not too large, and the edges are mostly directed from a higher layer to a lower layer. Then the vertices are ordered inside each layer in an attempt to minimize the number of edge crossings. Finally, each vertex is given a location (within its layer and respecting the ordering of that layer) so that edges are as straight as possible. The Sugiyama method involves a number

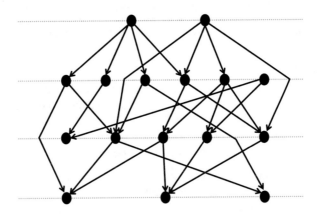

Fig. 2.30 A directed graph

of NP-hard combinatorial optimization problems, but each of these problems has heuristic solutions that work reasonably well in practice. The method has been refined and improved significantly since its original conception, most notably by Gansner et al. (1993).

2.5.2   Trees

Rooted trees are normally drawn with the root at the top of the screen, and parents above their children, as in Fig. 2.31a. Rooted trees are directed graphs, and the Sugiyama method described above can be used; however, simpler methods are available. A naïve algorithm to draw trees in this way is a simple exercise; however, a naive approach often leads to drawings that are too wide. Reingold and Tilford (1981) defined a linear-time algorithm that moves subtrees together to avoid excessive width. The Reingold–Tilford algorithm has been improved and extended many times (see, e.g., Buchheim et al. 2006). Unrooted trees can be drawn using energy-based methods. However, simple algorithms using drawing vertices on layers of concentric circles, as in Fig. 2.31b, are described in Battista et al. (1999).

2.5.3   Interaction

For large and complex graphs, interactive exploration is necessary. Interactive operations include:

- Computer-supported *filtering*. For example, the system may detect salient structural or semantic features, and filter out all vertices and edges that are not related to these features.

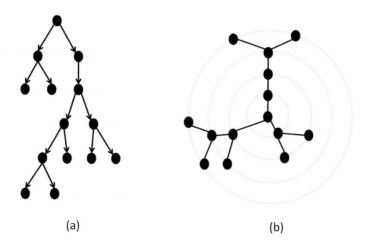

(a)   (b)

Fig. 2.31 (a) A rooted tree drawing; (b) an unrooted radial tree drawing

- *Zoom and pan.* If the user wishes to concentrate on a specific part of the graph, *focus+context* methods can be used in combination with a number of *pan* methods. These include screen-space methods such as fish-eye views (see, e.g., Furnas 2006) and slider bars, as well as graph-space methods (see, e.g., Eades et al. 1997).

2.5.4 More Metaphors

The node-link metaphor described in this chapter is most common, but other visual representations of graphs are used as well. Examples include the following:

- The *map metaphor* is used in Fig. 2.32a to show another picture of the graph in Fig. 2.2b. In this case, each vertex is represented by a region of the plane, and friendship between two people is represented as adjacency between regions. This metaphor has been developed extensively, from "treemaps" (Johnson and Shneiderman 1991) to "Gmaps" (Gansner et al. 2010).
- The *adjacency matrix metaphor*, illustrated in Fig. 2.32b, has been shown to be useful in some cases (Ghoniem et al. 2005).
- *Edge bundling.* For a dense and complex graph, edges can be bundled together as in Fig. 2.7. This method reduces edge clutter at the cost of reduced faithfulness; it seems to improve human understanding of global structural aspects of the graph.

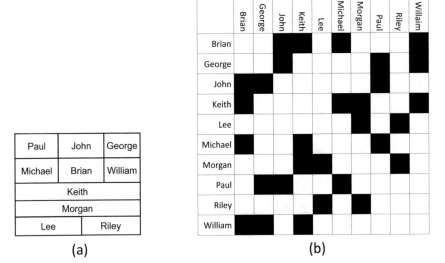

(a)   (b)

Fig. 2.32 Drawings of the graph in Fig. 2.2: (**a**) using the "map metaphor", (**b**) using the adjacency matrix metaphor

(a)   **(b)**   **(c)**

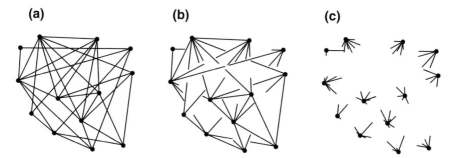

Fig. 2.33 Representation of edges as stubs to overcome readability problems due to crossings and clutter. Example as given in Bruckdorfer et al. (2012). (**a**) Original drawing with crossings. (**b**) Maximum stub size where both stubs have the same length. (**c**) Same as (**b**) with the additional constraint that the stub length to edge length ratio is the same for all edges—nearly nothing is left of the original edge lines

- *Edge stubs.* One interesting way to reduce edge clutter is by removing the parts of the edges where crossings occur, as in Fig. 2.33. For the case of directed graphs, Burch et al. (2012) and Bruckdorfer et al. (2012) show that for complex tasks the error rate increases with decreasing stub length, while for simple tasks (such as detection of the vertex with highest degree) the stub drawing can be beneficial in completion time and error rate.
- *Distortion.* Several approaches employ distortion as a way to reduce clutter for large graph visualization. One example is a focus+context technique that lays out the graph on the hyperbolic plane (Lamping et al. 1995). While this allows to put important objects in the focus center with a large part of the graph kept in the context area at the same time, readability might be reduced in areas with high distortion.

2.6   Concluding Remarks

The two graph drawing approaches described in Sects. 2.3 and 2.4 cover the main graph drawing algorithms in research and in practice. In addition to these general approaches, there exists a variety of algorithms to visualize specific classes of graphs like trees (Rusu 2013), dense graphs (Dwyer et al. 2014) or small-world and scale-free graphs (Nocaj et al. 2016; Jia et al. 2008). These algorithms exploit the characteristics of a graph class, and might be able to create improved visualizations for input instances from those classes. A large number of companies and organizations distribute graph drawing software. Some examples are as follows:

1. Tom Sawyer Software—This commercial enterprise (Tom Sawyer Software 2015) currently dominates the market for graphing software. Energy-based methods, orthogonal grid drawings, directed graph methods, and tree drawing

methods are included. The methods are packaged in many different ways to handle a variety of data sources.

2. OGDF—The Open Graph Drawing Framework (Chimani et al. 2013) is a self-contained open-source library of graph algorithms and data structures, freely available under the GPL at OGDF (2015). It is implemented in C++ and offers a wide variety of efficient graph drawing algorithm implementations, in particular covering planarization- and energy-based approaches. OGDF is maintained and used by several university research groups around the world.

3. Tulip—Tulip (TULIP 2015) is an information visualization framework that is freely available under the LPGL. It provides a Graphical User Interface and a C++ API. The GUI allows import of a variety of graph file formats and provides a range of layout algorithms.

4. WebCoLa—WebCoLa is an open-source JavaScript library for arranging HTML5 documents and diagrams using constraint-based optimization techniques. It supports interactive layout generation in a browser and works well with the well-known D3 library.

All of the above systems have one or more energy-based methods. In contrast, the topology-shape-metrics approach is seldom implemented in practical systems, despite significant attention from researchers and scientific evidence of readability. There are a number of possible reasons for the lack of commercial interest in the topology-shape-metrics approach: it is much more complex than the energy-based approach, and the approach does not seem to scale visually to larger graphs. Further research is needed to understand why energy-based methods are dominant in practice.

References

Angelini P, Binucci C, Lozzo GD, Didimo W, Grilli L, Montecchiani F, Patrignani M, Tollis IG (2013) Drawing non-planar graphs with crossing-free subgraphs. In: Wismath SK, Wolff A (eds) Graph drawing. In: 21st international symposium, GD 2013, Bordeaux, September 23–25, 2013. Revised selected papers. Lecture notes in computer science, vol 8242. Springer, Berlin, pp 292–303. http://dx.doi.org/10.1007/978-3-319-03841-4_26

Argyriou EN, Bekos MA, Kaufmann M, Symvonis A (2013) Geometric RAC simultaneous drawings of graphs. J Graph Algorithms Appl 17(1):11–34. http://dx.doi.org/10.7155/jgaa. 00282

Arikushi K, Fulek R, Keszegh B, Moric F, Tóth CD (2012) Graphs that admit right angle crossing drawings. Comput Geom 45(4):169–177. http://dx.doi.org/10.1016/j.comgeo.2011.11.008

Auer C, Brandenburg FJ, Gleißner A, Reislhuber J (2015) 1-Planarity of graphs with a rotation system. J Graph Algorithms Appl 19(1):67–86. http://dx.doi.org/10.7155/jgaa.00347

Barnard ST, Simon HD (1994) Fast multilevel implementation of recursive spectral bisection for partitioning unstructured problems. Concurrency Pract Experience 6(2):101–117. http://dx.doi. org/10.1002/cpe.4330060203

Bartel G, Gutwenger C, Klein K, Mutzel P (2010) An experimental evaluation of multilevel layout methods. In: Brandes U, Cornelsen S (eds) Graph drawing - 18th international symposium, GD 2010, Konstanz, September 21–24, 2010. Revised selected papers. Lecture notes in computer science, vol 6502, Springer, Berlin, pp 80–91. http://dx.doi.org/10.1007/978-3-642-18469-7_8

Batini C, Nardelli E, Tamassia R (1986) A layout algorithm for data flow diagrams. IEEE Trans Softw Eng 12(4):538–546. http://doi.ieeecomputersociety.org/10.1109/TSE.1986.6312901

Battista GD, Eades P, Tamassia R, Tollis IG (1999) Graph drawing: algorithms for the visualization of graphs. Prentice-Hall, Upper Saddle River

Biedl TC, Madden B, Tollis IG (1997) The three-phase method: a unified approach to orthogonal graph drawing. In: Battista GD (ed) (1997) Graph drawing. In: 5th international symposium, GD '97, Rome, September 18–20, 1997, Proceedings. Lecture notes in computer science, vol 1353. Springer, Berlin, pp 391–402. http://dx.doi.org/10.1007/3-540-63938-1_84

Böttger J, Schurade R, Jakobsen E, Schäfer A, Margulies D (2014) Connexel visualization: a software implementation of glyphs and edge-bundling for dense connectivity data using brainGL. Front Neurosci 8:15. https://doi.org/10.3389/fnins.2014.00015

Boyer JM, Myrvold WJ (2004a) On the cutting edge: simplified o (n) planarity by edge addition. J Graph Algorithms Appl 8(2):241–273

Boyer JM, Myrvold WJ (2004b) On the cutting edge: simplified o(n) planarity by edge addition. J Graph Algorithms Appl 8(2):241–273. http://jgaa.info/accepted/2004/BoyerMyrvold2004.8.3.pdf

Brandenburg FJ (2014) 1-visibility representations of 1-planar graphs. J Graph Algorithms Appl 18(3):421–438. http://dx.doi.org/10.7155/jgaa.00330

Bruckdorfer T, Cornelsen S, Gutwenger C, Kaufmann M, Montecchiani F, Nöllenburg M, Wolff A (2012) Progress on partial edge drawings. In: Didimo W, Patrignani M (eds) Graph drawing - 20th international symposium, GD 2012, Redmond, September 19–21, 2012. Revised selected papers. Lecture notes in computer science, vol 7704. Springer, Berlin, pp 67–78. http://dx.doi.org/10.1007/978-3-642-36763-2_7

Buchheim C, Jünger M, Leipert S (2006) Drawing rooted trees in linear time. Softw Pract Exper 36(6):651–665. http://dx.doi.org/10.1002/spe.713

Burch M, Vehlow C, Konevtsova N, Weiskopf D (2012) Evaluating partially drawn links for directed graph edges. In: Graph drawing, Springer, Berlin, pp 226–237

Chimani M, Klein K (2013) Shrinking the search space for clustered planarity. In: Graph drawing, Springer, Berlin, pp 90–101

Chimani M, Gutwenger C, Jünger M, Klau GW, Klein K, Mutzel P (2013) The open graph drawing framework (OGDF). In: Tamassia R (ed) (2013) Handbook on graph drawing and visualization. Chapman and Hall/CRC, Boca Raton, pp 543–569. https://www.crcpress.com/Handbook-of-Graph-Drawing-and-Visualization/Tamassia/9781584884125

Chimani M, Di Battista G, Frati F, Klein K (2014) Advances on testing c-planarity of embedded flat clustered graphs. In: Graph drawing, Springer Berlin, Heidelberg, pp 416–427

Chrobak M, Eppstein D (1991) Planar orientations with low out-degree and compaction of adjacency matrices. Theor Comput Sci 86(2):243–266

Cortese PF, Di Battista G (2005) Clustered planarity. In: Proceedings of the twenty-first annual symposium on computational geometry. ACM, New York SCG '05, pp 32–34. http://doi.acm.org/10.1145/1064092.1064093

Czauderna T, Klukas C, Schreiber F (2010) Editing, validating, and translating of SBGN maps. Bioinformatics 26(18):2340–2341

Dahlhaus E (1998) A linear time algorithm to recognize clustered graphs and its parallelization. In: Lucchesi CL, Moura AV (eds) LATIN '98: theoretical Informatics. Third Latin American symposium, Campinas, April, 20–24, 1998, Proceedings. Lecture notes in computer science, vol 1380. Springer, Berlin, pp 239–248. http://dx.doi.org/10.1007/BFb0054325

Didimo W, Eades P, Liotta G (2009) Drawing graphs with right angle crossings. In: Dehne FKHA, Gavrilova ML, Sack J, Tóth CD (eds) Algorithms and data structures. 11th International Symposium, WADS 2009, Banff, August 21–23, 2009. Proceedings. Lecture notes in computer science, vol 5664, Springer, Berlin, pp 206–217. http://dx.doi.org/10.1007/978-3-642-03367-4_19

Dwyer T (2009) Scalable, versatile and simple constrained graph layout. Comput Graph Forum 28(3):991–998

Dwyer T, Mears C, Morgan K, Niven T, Marriott K, Wallace M (2014) Improved optimal and approximate power graph compression for clearer visualisation of dense graphs. In: Fujishiro I, Brandes U, Hagen H, Takahashi S (eds) IEEE pacific visualization symposium, PacificVis 2014, Yokohama, March 4–7, 2014. IEEE Computer Society, New York, pp 105–112. http://dx.doi.org/10.1109/PacificVis.2014.46

Eades P (1984) A heuristics for graph drawing. Congressus numerantium 42:146–160

Eades P, Liotta G (2013) Right angle crossing graphs and 1-planarity. Discrete Appl Math 161(7–8):961–969. http://dx.doi.org/10.1016/j.dam.2012.11.019

Eades P, Cohen RF, Huang ML (1997) Online animated graph drawing for web navigation. In: Battista GD (ed) (1997) Graph drawing. In: 5th international symposium, GD '97, Rome, September 18–20, 1997, Proceedings. Lecture notes in computer science, vol 1353. Springer, Berlin, pp 330–335. http://dx.doi.org/10.1007/3-540-63938-1_77

Eades P, Feng Q, Nagamochi H (1999) Drawing clustered graphs on an orthogonal grid. J Graph Algorithms Appl 3(4):3–29. http://www.cs.brown.edu/publications/jgaa/accepted/99/EadesFengNagamochi99.3.4.pdf

Eades P, Hong S, Katoh N, Liotta G, Schweitzer P, Suzuki Y (2013) A linear time algorithm for testing maximal 1-planarity of graphs with a rotation system. Theor Comput Sci 513:65–76. http://dx.doi.org/10.1016/j.tcs.2013.09.029

Eades P, Hong S, Klein K, Nguyen A (2015) Shape-based quality metrics for large graph visualization. In: Giacomo ED, Lubiw A (eds) Graph drawing and network visualization - 23rd international symposium, GD 2015, Los Angeles, September 24–26, 2015. Revised selected papers. Lecture notes in computer science, vol 9411. Springer, Berlin, pp 502–514. http://dx.doi.org/10.1007/978-3-319-27261-0_41

Feng Q, Cohen RF, Eades P (1995) Planarity for clustered graphs. In: Spirakis PG (ed) Algorithms - ESA '95, Third annual European symposium, Corfu, September 25–27, 1995, Proceedings. Lecture notes in computer science, Springer, Berlin, vol 979, pp 213–226. http://dx.doi.org/10.1007/3-540-60313-1_145

Fößmeier U, Kaufmann M (1995) Drawing high degree graphs with low bend numbers. In: Brandenburg F (ed) (1996) Graph drawing. In: Symposium on graph drawing, GD '95, Passau, September 20–22, 1995, Proceedings. Lecture notes in computer science, vol 1027. Springer, Berlin, pp 254–266. http://dx.doi.org/10.1007/BFb0021809

Fruchterman TMJ, Reingold EM (1991) Graph drawing by force-directed placement. Softw Pract Exper 21(11):1129–1164. http://dx.doi.org/10.1002/spe.4380211102

Furnas GW (2006) A fisheye follow-up: further reflections on focus + context. In: Grinter RE, Rodden T, Aoki PM, Cutrell E, Jeffries R, Olson GM (eds) Proceedings of the 2006 conference on human factors in computing systems, CHI 2006, Montréal, Québec, April 22–27, 2006. ACM, New York, pp 999–1008. http://doi.acm.org/10.1145/1124772.1124921

Gajer P, Goodrich MT, Kobourov SG (2004) A multi-dimensional approach to force-directed layouts of large graphs. Comput Geom 29(1):3–18. http://dx.doi.org/10.1016/j.comgeo.2004.03.014

Gansner ER, Koutsofios E, North SC, Vo K (1993) A technique for drawing directed graphs. IEEE Trans Softw Eng 19(3):214–230. http://dx.doi.org/10.1109/32.221135

Gansner ER, Koren Y, North SC (2004) Graph drawing by stress majorization. In: Pach J (ed) (2004) Graph drawing. In: 12th international symposium, GD 2004, New York, September 29 - October 2, 2004. Revised selected papers. Lecture notes in computer science, vol 3383. Springer, Berlin, pp 239–250. http://dx.doi.org/10.1007/978-3-540-31843-9_25

Gansner ER, Hu Y, Kobourov SG (2010) Gmap: visualizing graphs and clusters as maps. In: IEEE pacific visualization symposium pacificVis 2010, Taipei, March 2–5, 2010. IEEE, New York, pp 201–208. http://dx.doi.org/10.1109/PACIFICVIS.2010.5429590

Gansner ER, Hu Y, Krishnan S (2013a) Coast: a convex optimization approach to stress-based embedding. In: Wismath SK, Wolff A (eds) Graph drawing. In: 21st international symposium, GD 2013, Bordeaux, September 23–25, 2013. Revised selected papers. Lecture notes in computer science, vol 8242. Springer, Berlin. http://dx.doi.org/10.1007/978-3-319-03841-4 pp 268–279. http://dx.doi.org/10.1007/978-3-319-03841-4

Gansner ER, Hu Y, North SC (2013b) A maxent-stress model for graph layout. IEEE Trans Vis Comput Graph 19(6):927–940. http://doi.ieeecomputersociety.org/10.1109/TVCG.2012.299

Gansner ER, Hu Y, North SC (2013c) A maxent-stress model for graph layout. IEEE Trans Vis Comput Graph 19(6):927–940. http://doi.ieeecomputersociety.org/10.1109/TVCG.2012.299

Garey MR, Johnson DS (1979) Computers and intractability: a guide to the theory of NP-completeness. W. H. Freeman, New York

Garg A, Tamassia R (1996) A new minimum cost flow algorithm with applications to graph drawing. In: North SC (ed) Graph drawing, Symposium on graph drawing, GD '96, Berkeley, September 18–20, Proceedings. Lecture notes in computer science, vol 1190. Springer, Berlin, pp 201–216. http://dx.doi.org/10.1007/3-540-62495-3_49

Ghoniem M, Fekete J, Castagliola P (2005) On the readability of graphs using node-link and matrix-based representations: a controlled experiment and statistical analysis. Inform Vis 4(2):114–135. http://dx.doi.org/10.1057/palgrave.ivs.9500092

Giacomo ED, Liotta G, Montecchiani F (2014) Drawing outer 1-planar graphs with few slopes. In: Duncan CA, Symvonis A (eds) Graph drawing - 22nd international symposium, GD 2014, Würzburg, September 24–26, 2014. Revised selected papers. Lecture notes in computer science, vol 8871. Springer, Berlin, pp 174–185. http://dx.doi.org/10.1007/978-3-662-45803-7_15

Gutwenger C, Jünger M, Leipert S, Mutzel P, Percan M, Weiskircher R (2002) Advances in c-planarity testing of clustered graphs. In: Goodrich M, Kobourov S (eds) Graph drawing. Lecture notes in computer science, vol 2528. Springer, Berlin, Heidelberg, pp 220–236. http://dx.doi.org/10.1007/3-540-36151-0_21

Gutwenger C, Jünger M, Klein K, Kupke J, Leipert S, Mutzel P (2003) A new approach for visualizing uml class diagrams. In: Proceedings of the 2003 ACM symposium on software visualization. ACM, New York, pp 179–188

Hachul S, Jünger M (2004) Drawing large graphs with a potential-field-based multilevel algorithm. In: Pach J (ed) (2004) Graph drawing. In: 12th international symposium, GD 2004, New York, September 29 - October 2, 2004. Revised selected papers. Lecture notes in computer science, vol 3383. Springer, Berlin, pp 285–295

Hachul S, Jünger M (2007) Large-graph layout algorithms at work: an experimental study. J Graph Algorithms Appl 11(2):345–369

Hadany R, Harel D (2001) A multi-scale algorithm for drawing graphs nicely. Discrete Applied Mathematics 113(1):3–21. http://dx.doi.org/10.1016/S0166-218X(00)00389-9

Hall K (1970) An r-dimensional quadratic placement algorithm. Management Science 17:219–229

Harel D, Koren Y (2002) A fast multi-scale method for drawing large graphs. J Graph Algorithms Appl 6(3):179–202, http://www.cs.brown.edu/publications/jgaa/accepted/2002/HarelKoren2002.6.3.pdf

Harel D, Koren Y (2004) Graph drawing by high-dimensional embedding. J Graph Algorithms Appl 8(2):195–214, http://jgaa.info/accepted/2004/HarelKoren2004.8.2.pdf

Holten D, van Wijk JJ (2009) Force-directed edge bundling for graph visualization. Comput Graph Forum 28(3):983–990. http://dx.doi.org/10.1111/j.1467-8659.2009.01450.x

Hopcroft JE, Tarjan RE (1974) Efficient planarity testing. J ACM 21(4):549–568. http://doi.acm.org/10.1145/321850.321852

Hu Y, Shi L (2015) Visualizing large graphs. Wiley Interdisciplinary Reviews: Computational Statistics 7(2):115–136. http://dx.doi.org/10.1002/wics.1343

Huang W, Eades P, Hong S (2014) Larger crossing angles make graphs easier to read. J Vis Lang Comput 25(4):452–465. http://dx.doi.org/10.1016/j.jvlc.2014.03.001

Jelínková E, Kára J, Kratochvíl J, Pergel M, Suchý O, Vyskocil T (2009) Clustered planarity: Small clusters in cycles and Eulerian graphs. J Graph Algorithms Appl 13(3):379–422, http://jgaa.info/accepted/2009/Jelinkova+2009.13.3.pdf

Jia Y, Hoberock J, Garland M, Hart J (2008) On the visualization of social and other scale-free networks. IEEE Transactions on Visualization and Computer Graphics 14(6):1285–1292. https://doi.org/10.1109/TVCG.2008.151

Johnson B, Shneiderman B (1991) Tree maps: A space-filling approach to the visualization of hierarchical information structures. In: IEEE Visualization, pp 284–291. http://dx.doi.org/10.1109/VISUAL.1991.175815

Jünger M, Mutzel P (1994) The polyhedral approach to the maximum planar subgraph problem: New chances for related problems. In: Tamassia R, Tollis IG (eds) (1995) Graph drawing. In: DIMACS international workshop, GD '94, Princeton, October 10–12, 1994, Proceedings. Lecture notes in computer science, vol 894. Springer, Berlin, pp 119–130. http://dx.doi.org/10.1007/3-540-58950-3_363

Jünger M, Mutzel P (1996) Maximum planar subgraphs and nice embeddings: Practical layout tools. Algorithmica 16(1):33–59. http://dx.doi.org/10.1007/BF02086607

Jünger M, Leipert S, Mutzel P (1998) A note on computing a maximal planar subgraph using pq-trees. IEEE Trans on CAD of Integrated Circuits and Systems 17(7):609–612. http://doi.ieeecomputersociety.org/10.1109/43.709399

Kamada T, Kawai S (1989) An algorithm for drawing general undirected graphs. Inf Process Lett 31(1):7–15. http://dx.doi.org/10.1016/0020-0190(89)90102-6

Karypis G, Kumar V (1995) Analysis of multilevel graph partitioning. In: Karin S (ed) Proceedings supercomputing '95, San Diego, December 4-8, 1995. IEEE Computer Society/ACM, New York, p 29. http://doi.acm.org/10.1145/224170.224229

Khoury M, Hu Y, Krishnan S, Scheidegger CE (2012) Drawing large graphs by low-rank stress majorization. Comput Graph Forum 31(3):975–984. http://dx.doi.org/10.1111/j.1467-8659.2012.03090.x

Kobourov SG (2013) Force-directed drawing algorithms. Handbook of graph drawing and visualization, pp 383–408

Kobourov SG, Pupyrev S, Saket B (2014) Are crossings important for drawing large graphs? In: Graph drawing. Springer, Berlin, pp 234–245

Koren Y, Carmel L, Harel D (2002) ACE: a fast multiscale eigenvectors computation for drawing huge graphs. In: Wong PC, Andrews K (eds) 2002 IEEE symposium on information visualization (InfoVis 2002), 27 October–1 November 2002, Boston. IEEE Computer Society, New York, pp 137–144. http://dx.doi.org/10.1109/INFVIS.2002.1173159

Kuratowski K (1930) Sur le problème des courbes gauches en topologie. Fund Math 15:271–283

Lamping J, Rao R, Pirolli P (1995) A focus+context technique based on hyperbolic geometry for visualizing large hierarchies. In: Proceedings of the SIGCHI conference on human factors in computing systems, CHI '95. ACM Press/Addison-Wesley, New York, pp 401–408. http://dx.doi.org/10.1145/223904.223956

Le Novère N, Hucka M, Mi H, Moodie S, Schreiber F, Sorokin A, Demir E, Wegner K, Aladjem M, Wimalaratne SM, Bergman FT, Gauges R, Ghazal P, Kawaji H, Li L, Matsuoka Y, Villéger A, Boyd SE, Calzone L, Courtot M, Dogrusoz U, Freeman T, Funahashi A, Ghosh S, Jouraku A, Kim S, Kolpakov F, Luna A, Sahle S, Schmidt E, Watterson S, Wu G, Goryanin I, Kell DB, Sander C, Sauro H, Snoep JL, Kohn K, Kitano H (2009) The systems biology graphical notation. Nat Biotechnol 27:735–741

Lin T, Eades P (1994) Integration of declarative and algorithmic approaches for layout creation. In: Tamassia R, Tollis IG (eds) (1995) Graph drawing. In: DIMACS international workshop, GD '94, Princeton, October 10–12, 1994, Proceedings. Lecture notes in computer science, vol 894. Springer, Berlin, pp 376–387. http://dx.doi.org/10.1007/3-540-58950-3_392

Mehlhorn K, Mutzel P (1996) On the embedding phase of the Hopcroft and Tarjan planarity testing algorithm. Algorithmica 16(2):233–242. http://dx.doi.org/10.1007/BF01940648

Misue K, Eades P, Lai W, Sugiyama K (1995) Layout adjustment and the mental map. J Vis Lang Comput 6(2):183–210. http://dx.doi.org/10.1006/jvlc.1995.1010

Monash University (2015) WebCoLa – constraint-based layout in the browser. http://marvl.
    infotech.monash.edu/webcola/

Newbery FJ (1989) Edge concentration: a method for clustering directed graphs. In: SCM,
    pp 76–85

Nguyen QH, Eades P, Hong S (2013) On the faithfulness of graph visualizations. In: Carpendale
    S, Chen W, Hong S (eds) IEEE pacific visualization symposium, PacificVis 2013, February
    27 2013-March 1, 2013, Sydney. IEEE, New York, pp 209–216. http://dx.doi.org/10.1109/
    PacificVis.2013.6596147

Nocaj A, Ortmann M, Brandes U (2016) Adaptive disentanglement based on local clustering in
    small-world network visualization. IEEE Trans Vis Comput Graph. http://dx.doi.org/10.1109/
    TVCG.2016.2534559

OGDF (2015) The open graph drawing framework. http://www.ogdf.net

Purchase HC (2002) Metrics for graph drawing aesthetics. J Vis Lang Comput 13(5):501–516.
    http://dx.doi.org/10.1006/jvlc.2002.0232

Purchase HC, Cohen RF, James MI (1995) Validating graph drawing aesthetics. In: Brandenburg F
    (ed) (1996) Graph drawing. In: Symposium on graph drawing, GD '95, Passau, September 20–
    22, 1995, Proceedings. Lecture notes in computer science, vol 1027. Springer, Berlin, pp 435–
    446. http://dx.doi.org/10.1007/BFb0021827

Quigley A, Eades P (2001) Fade: graph drawing, clustering, and visual abstraction. In: Graph
    drawing. Springer, Berlin, Heidelberg, pp 197–210

Reingold EM, Tilford JS (1981) Tidier drawings of trees. IEEE Trans Softw Eng 7(2):223–228.
    http://dx.doi.org/10.1109/TSE.1981.234519

Rohn H, Junker A, Hartmann A, Grafahrend-Belau E, Treutler H, Klapperstuck M, Czauderna T,
    Klukas C, Schreiber F (2012) Vanted v2: a framework for systems biology applications. BMC
    Syst Biol 6:139

Rüegg U, Kieffer S, Dwyer T, Marriott K, Wybrow M (2014) Stress-minimizing orthogonal layout
    of data flow diagrams with ports. In: Graph drawing. Springer, Berlin, pp 319–330

Rusu A (2013) Three drawing algorithms. In: Tamassia R (ed) (2013) Handbook on graph drawing
    and visualization. Chapman and Hall/CRC, Boca Raton, pp 155–192. https://www.crcpress.
    com/Handbook-of-Graph-Drawing-and-Visualization/Tamassia/9781584884125

Schreiber F, Dwyer T, Marriott K, Wybrow M (2009) A generic algorithm for layout of biological
    networks. BMC Bioinform 10:375

Shih W, Hsu W (1999) A new planarity test. Theor Comput Sci 223(1–2):179–191. http://dx.doi.
    org/10.1016/S0304-3975(98)00120-0

Sugiyama K, Tagawa S, Toda M (1981) Methods for visual understanding of hierarchical system
    structures. IEEE Trans Syst Man Cybern 11(2):109–125. http://dx.doi.org/10.1109/TSMC.
    1981.4308636

Sultana S, Rahman MS, Roy A, Tairin S (2014) Bar 1-visibility drawings of 1-planar graphs.
    In: Gupta P, Zaroliagis CD (eds) Applied algorithms - first international conference, ICAA
    2014, Kolkata, January 13–15, 2014. Proceedings. Lecture notes in computer science, vol 8321.
    Springer, Berlin, pp 62–76. http://dx.doi.org/10.1007/978-3-319-04126-1_6

Tamassia R (1987) On embedding a graph in the grid with the minimum number of bends. SIAM
    J Comput 16(3):421–444. http://dx.doi.org/10.1137/0216030

Tamassia R, Tollis IG (1986) Algorithms for visibility representations of planar graphs. In: Monien
    B, Vidal-Naquet G (eds) STACS 86, 3rd annual symposium on theoretical aspects of computer
    science, Orsay, January 16–18, 1986, Proceedings. Lecture notes in computer science, vol 210.
    Springer, Berlin, pp 130–141. http://dx.doi.org/10.1007/3-540-16078-7_71

Tamassia R, Battista GD, Batini C (1988) Automatic graph drawing and readability of diagrams.
    IEEE Trans Syst Man Cybern 18(1):61–79. http://dx.doi.org/10.1109/21.87055

Tom Sawyer Software (2015) Tom sawyer toolkit. https://www.tomsawyer.com/

Torgerson WS (1952) Multidimensional scaling: I. theory and method. Psychometrika 17(4):401–
    419. http://dx.doi.org/10.1007/BF02288916

Tufte ER (1992) The visual display of quantitative information. Graphics Press, Cheshire

TULIP (2015) The Tulip framework. tulip.labri.fr

Tutte WT (1960) Convex representations of graphs. Proc Lond Math Soc 10:304–320

Tutte WT (1963) How to draw a graph. Proc Lond Math Soc 13:743–767

University of Florida (2015) The university of Florida sparse matrix collection. http://www.cise.ufl.edu/research/sparse/matrices/

Walshaw C (2003) A multilevel algorithm for force-directed graph-drawing. J Graph Algorithms Appl 7(3):253–285. http://www.cs.brown.edu/publications/jgaa/accepted/2003/Walshaw2003.7.3.pdf

Ware C, Purchase HC, Colpoys L, McGill M (2002) Cognitive measurements of graph aesthetics. Inf. Vis. 1(2):103–110. http://dx.doi.org/10.1057/palgrave.ivs.9500013

Yoghourdjian V, Dwyer T, Gange G, Kieffer S, Klein K, Marriott K (2016) High-quality ultra-compact grid layout of grouped networks. IEEE Trans Vis Comput Graph 22(1):339–348. http://doi.ieeecomputersociety.org/10.1109/TVCG.2015.2467251

Yunis E, Yokota R, Ahmadia AJ (2012) Scalable force directed graph layout algorithms using fast multipole methods. In: Bader M, Bungartz H, Grigoras D, Mehl M, Mundani R, Potolea R (eds) 11th international symposium on parallel and distributed computing, ISPDC 2012, Munich, June 25–29, 2012. IEEE Computer Society, New York, pp 180–187. http://dx.doi.org/10.1109/ISPDC.2012.32

Chapter 3
gLabTrie: A Data Structure for Motif Discovery with Constraints

Misael Mongiovì, Giovanni Micale, Alfredo Ferro, Rosalba Giugno, Alfredo Pulvirenti, and Dennis Shasha

Abstract Motif discovery is the problem of finding subgraphs of a network that appear surprisingly often. Each such subgraph may indicate a small-scale interaction feature in applications ranging from a genomic interaction network, a significant relationship involving rock musicians, or any other application that can be represented as a network. We look at the problem of constrained search for motifs based on labels (e.g. gene ontology, musician type to continue our example from above). This chapter presents a brief review of the state of the art in motif finding and then extends the gTrie data structure from Ribeiro and Silva (Data Min Knowl Discov 28(2):337–377, 2014b) to support labels. Experiments validate the usefulness of our structure for small subgraphs, showing that we recoup the cost of the index after only a handful of queries.

3.1 The Problem and Its Motivation

A motif in a graph is a subgraph that appears statistically significantly often. Frequently occurring motifs may have practical significance. One familiar example is the ubiquity of feedback networks underlying homeostasis in biological, natural, and even economic systems. Motifs can also be useful in engineering disciplines such as synthetic biology. Kurata et al. (2014) use the frequent motifs found in biological networks as a library for synthetic biology. In fact, Kurata et al. pointed out that there are often motifs that behave as a single node in a larger network motif,

M. Mongiovì · G. Micale
Department of Maths and Computer Science, University of Catania, Catania, Italy
e-mail: mongiovi@dmi.unict.it; gmicale@dmi.unict.it

A. Ferro · R. Giugno · A. Pulvirenti
Department of Clinical and Experimental Medicine, University of Catania, Catania, Italy
e-mail: ferro@dmi.unict.it; giugno@dmi.unict.it; pulvirenti@dmi.unict.it

D. Shasha (✉)
Courant Institute of Mathematical Science, New York University, New York, NY, USA
e-mail: shasha@courant.nyu.edu

© Springer International Publishing AG, part of Springer Nature 2018
G. Fletcher et al. (eds.), *Graph Data Management*, Data-Centric Systems and Applications, https://doi.org/10.1007/978-3-319-96193-4_3

just as an AND gate in an electronic circuit built out of transistors and resistors acts as a single node in a logic diagram. So, there may be motifs at different levels of abstraction. For the purposes of our chapter, we will take the usefulness of motifs for granted and talk about how to discover such motifs efficiently.

Further, we will be particularly concerned with graphs whose vertices have labels. A constrained labeled motif query is to find a statistically significant motif satisfying some constraint on the labels.

In the sequel, we will define our notion of statistical significance, but informally, this will entail a simulation of the following process: (1) Find many random variations of the input graph G where each random variation preserves the degree counts of each node in G and preserves the number of edges linking nodes having each pair of labels. (2) See how often a labeled topological structure of interest is found in those random graphs. If infrequently, then the labeled topological structure is significant in G and constitutes a motif.

The computational challenge in motif finding is that the number of possible subgraphs could, depending on the graph, grow exponentially with the size of the subgraph. For sparser graphs, the growth may be less dramatic, but still rapid.

For that reason, we use data structures to make this fast. Our work builds directly on the gTrie data structure developed by Ribeiro and Silva (2012) which is why we call our structure gLabTrie.

This chapter begins with a discussion of the data structure and algorithm we will use. We then follow with a discussion of how to find rare structures. Finally, we give an experimental evaluation of our structure and algorithms.

3.2 gLabTrie Structure

3.2.1 Preliminaries

For simplicity, our discussion will center on undirected graphs, although our method works with directed graphs as well. Given a graph G, we denote by V_G its set of vertices, by E_G its set of edges, by L_G its alphabet of labels, and by l_G a function that assigns a label to each vertex. We also write $G = (V_G, E_G, L_G, l_G)$. A subgraph G' of a graph G (denoted by $G' \subseteq G$) is a graph that contain, a subset of vertices $V_{G'} \subseteq V_G$ of G and all edges of G whose endpoints are both in $V_{G'}$.

An *isomorphism* between two graphs G_1 and G_2 is a one-to-one mapping $\varphi : V_{G_1} \rightarrow V_{G_2}$ between vertices, which preserves the structure, i.e., $(u, v) \in G_1 \Leftrightarrow (\varphi(u), \varphi(v)) \in G_2$, and the labels, i.e., $l_G(u) = l_G(\varphi(u))$. If there is at least an isomorphism between G_1 and G_2, we say that they are *isomorphic* and write $G_1 \sim G_2$. An *automorphism* in G is an isomorphism between G and itself. Every graph admits at least one automorphism (where each vertex corresponds to itself). Typically, a graph can have many automorphisms. We abuse the notation and write $\varphi(G)$, with $G \subseteq G_1$ to denote the subgraph of G_2 that corresponds to G

according to φ (i.e. the subgraph composed of vertices $\varphi(v)$ with $v \in G$ and edges $(\varphi(v_1), \varphi(v_2))$ with $(v_1, v_2) \in G$).

In what follows, we use the terms *input network* (denoted by $\mathbb{G}$), *topologies* (denoted by G), i.e., unlabeled graphs that represent motif structures and *topology instances* (denoted by g), i.e., subgraphs of $\mathbb{G}$ that accommodate certain topologies. A *labeled topology* is an undirected (vertex-) labeled connected graph G. An *unlabeled topology* is a labeled topology stripped of its labels. A labeled topology that occurs frequently in G is also called *motif*. An *occurrence* g of a topology G is a connected subgraph of $\mathbb{G}$ that is isomorphic to G. So, a given topology may have zero, one, or more occurrences in a graph.

Checking whether two (labeled or unlabeled) topologies are isomorphic is an expensive task that requires finding an isomorphism between the topologies or proving that no isomorphism exists. In motif discovery, this operation has to be performed frequently to map a topology to the network subgraphs that conform to that topology. To simplify this operation, we map a graph to its *canonical form*, i.e., a string that uniquely identifies a topology and is invariant with respect to isomorphism. In other words, two isomorphic graphs should have the same canonical form, while two graphs that are not isomorphic should have different canonical forms. Computing the canonical form of a graph may be expensive, but once it is computed, the isomorphism check entails simple string comparison.

An easy way to find a canonical form for an unlabeled subgraph is to consider all possible adjacency matrices of that subgraph (by reordering vertices in all possible ways), linearize them into strings (by putting all rows of an adjacency matrix contiguously in a unique line) and considering the smallest string (with respect to a lexicographic order). This simple approach guarantees invariance with respect to isomorphism since two isomorphic graphs have the same adjacency matrix except for their rows/columns order. The approach can be generalized to labeled topologies by including the sequence of labels in the string. Since enumerating all possible vertex orders is impractical, more efficient methods have been defined. A widely used method is nauty (McKay 1981).

The canonical form of a graph is associated to a *canonical order* of vertices, i.e., the order of vertices that produces it. Note that a canonical form may be associated with more than one canonical order since a graph may have several automorphisms.

3.2.2 Problem Definition

We aim to support *label-based queries* in which the user specifies a set of constraints and the system returns all topologies that satisfy the constraints. In our framework, a user specifies a frequency threshold, a p-value threshold, and a bag (multiset) of labels that the motifs must contain. An example query would be: "Give me all labeled topologies of size k that have at least two A labels and one B label, occur at least f times and have a p-value smaller than p." We also want the query processing to be fast, so when a user is not satisfied with the response, he or she can change the constraints and quickly get a new response. We accept a slow (but still reasonable) offline *preprocessing* step in exchange for fast *query processing*.

Formally, we define a *label-based query* (more simply a query) as a quadruple $Q = (C, k, f, p)$, where C is a bag of labels (a bag, also called a multiset, is similar to a set, but an element may occur more than once), k is the requested size of motifs, f is a frequency threshold, and p is a p-value threshold.

Definition 3.1 (Label-Based Query Processing) Given a network $\mathbb{G}$ and a query $Q = (C, k, f, p)$, find all labeled topologies T with number of vertices (size) k, whose number of occurrences in $\mathbb{G}$ is at least f and whose p-value is no more than p.

We solve the defined problem in two steps. During an offline *preprocessing* phase, we census the input network to find all labeled motifs up to a certain size K, and organize them in a suitable data structure (that we call the *TopoIndex*). Later, during the online *query processing* phase, we probe the TopoIndex to efficiently retrieve motifs that satisfy the query constraints.

In the remaining part of this section, we describe how we extend existing approaches to support labeled motifs and the data structure used for quickly processing queries. Since our approach has been implemented on top of G-Trie, we first give an overview of G-Trie and our subsequent description will refer to it. However, our approach is general in that it can be applied on other network-centric algorithms for motif discovery.

3.2.3 G-Trie Method for Unlabeled Motif Discovery

The main data structure of a network-centric method for motif discovery is a key-value map (hash table or search tree) that associates each unlabeled topology (up to a certain size) to a counter. Unlabeled topologies may be represented by their canonical form, so that the isomorphic check is efficient. G-Trie (Ribeiro and Silva 2014b) generalizes tries to graphs. A gTrie organizes a set of unlabeled topologies in a multiway tree in such a way that subgraphs correspond to ancestors. An example of gTrie that stores all unlabeled topologies of size up to four vertices is given in Fig. 3.1.

Each node[1] of the gTrie stores information associated to the corresponding topology, typically a counter (not shown in the figure). A gTrie can be seen as a map that associates topologies to counters (similar in principle to a hash table or a binary tree).

---

[1] We use the term node to refer to parts of our data structures and vertex to talk about the graphs in which we find patterns.

Fig. 3.1 Example of a gTrie with $K = 4$. The data structure stores all unlabeled topologies with up to 4 vertices. A similar, more detailed example can be found in Ribeiro and Silva (2014b)

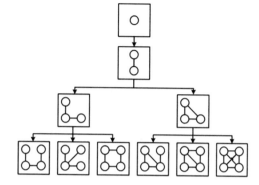

Algorithm 1: Network-centric algorithm for unlabeled motif discovery: first find topologies in the input network that meet the frequency threshold, then compare the number of occurrences with the number of occurrences in each of a set of random graphs to evaluate the p-value of each such topology

---

Require: network, size K, frequency threshold f, p-value threshold p, number of randomizations r {returns the set of motifs with frequency $\geq f$ and p-value $\leq p$}
initialize $gTrie$ with depth K
call $census(network, gTrie)$
initialize map_count
for $i = 0 \ldots r$ **do**
  $rand\_net = randomize(network)$
  initialize $gTrie\_rand$ with depth K
  call $census(rand\_net, gTrie\_rand)$
  for all $t \in topologies(gTrie\_rand)$ **do**
    if $gTrie\_rand[t] \geq gTrie[t]$ **then**
      $map\_count[t] = map\_count[t] + 1$
    end if
  end for
end for
for all $t \in keys(map\_count)$ **do**
  $pval = map\_count[t]/r$
  if $gTrie[t] \geq f$ and $pval \leq p$ **then**
    output t, $gTrie[t]$, $pval$
  end if
end for

To compute p-values, the GTrie system counts the number of occurrences of all unlabeled topologies in the input network and compares them with the corresponding number of occurrences in random networks with similar properties. The overall algorithm is in the figure marked Algorithm 1.

First a gTrie with all unlabeled topologies up to size K is built in the input network. Then the core procedure, $census()$, which takes as input a network and fills the gTrie[2] with the correct counting, is called. This procedure enumerates all subgraphs of the network one by one and increases the counter of the corresponding topology. Then, a map of counters (*map_count*) is initialized. This map is a hash table that associates topologies (more precisely canonical forms of topologies) to counters and is used to store the number of random networks in which a given topology occurs more than in the input network. Next, a number of randomizations of the input network are computed and *census*() is executed on each of them. For every topology found, if its number of occurrence is greater than the one in the input network, its counter is increased. Function *topologies*(*gTrie*) returns all topologies stored in *gTrie* while *gTrie*[*t*] refers to the counter associated with topology *t* in *gTrie*. At the end, frequencies and *p*-values are computed and all topologies that satisfy the input constraints are returned. In the next paragraphs we give more details about the core procedure, *census*(). Further details on the other parts can be found in Ribeiro and Silva (2014b).

The algorithm for graph census (procedure *census*()) is detailed in Algorithm 2. The algorithm is based on the recursive procedure *Match* that matches paths of the gTrie with all possible subgraph of the input network. At the beginning, the procedure *Match* is called on the root of the gTrie and with an empty subgraph ($V_{used} = \emptyset$). The procedure picks one vertex at a time and starts to grow a subgraph from that vertex. Every time a new child of a gTrie node is explored, all neighbors of previously taken vertices ($N(V_{used})$) are considered and, if matchable, associated with the current node and added to the current subgraph (V_{used}). When a leaf node is considered, the node counter is increased. This means that a new subgraph isomorphic to the topology associated to that node was found.

To perform a correct counting, every subgraph should be counted exactly once. Without symmetry breaking conditions, the *Match* procedure would find some subgraphs multiple times. Indeed, if a subgraph has more than one automorphism (isomorphism between it and itself) there are multiple ways to obtain it. For instance, consider a network that contains a triangle with vertex ids 1, 2, and 3. The enumeration would produce the same triangle six times with the following sequences: $(1, 2, 3)$, $(1, 3, 2)$, $(2, 1, 3)$, $(2, 3, 1)$, $(3, 1, 2)$, and $(3, 2, 1)$. Although multiple copies may be discarded by a post-processing step, this would require storing all subgraphs, which would be expensive for large subgraphs. Instead, the census algorithm considers a carefully designed set of symmetry-breaking conditions that guarantees that each subgraph is enumerated exactly once. In the specific example, the breaking conditions impose that the first vertex's identifier must be smaller than the second one's, and the second vertex's id must be smaller than the third one's. Thus, only $(1, 2, 3)$ would be a valid sequence of vertices for the triangle. Details on how symmetry-breaking conditions are computed are given in Ribeiro and Silva (2014b).

---

[2]In general the overall algorithm can work with any data structure that associates keys to values (e.g. hash tables) in place of gTrie. Keys are canonical forms of topologies, while values are counters.

Algorithm 2: Census algorithm for unlabeled motif discovery

Require: network, gTrie {returns the gTrie filled with the number of occurrences of each topology.}
$Match(gTrie.root, \emptyset)$
return $gTrie$

Procedure $Match(node, V_{used})$
if $V_{used} = \emptyset$ **then**
    $V_{cand} \leftarrow V(network)$
else
    $V_{cand} \leftarrow \{v \in N(V_{used}) : v$ satisfies symmetry breaking conditions$\}$
end if
$V \leftarrow \emptyset$
for all $v \in V_{cand}$ **do**
    if v is connected with V_{used} as defined in $node$ **then**
        $V \leftarrow V \cup \{v\}$
    end if
end for
for all $v \in V$ **do**
    if $isLeaf(node)$ **then**
        $node.counter + = 1$
    end if
    for all children c of $node$ **do**
        $Match(c, V_{used} \cup \{v\})$
    end for
end for
End Procedure

3.2.4 gLabTrie Data Structure for Labeled Motif Discovery

A naive extension for handling labeled networks would consist in incorporating labels into the gTrie nodes. A node would represent a labeled topology as opposed to an unlabeled topology. However, this approach would cause an explosion of the number of gTrie nodes as the number of labels grows. Each node has to maintain both connectivity and label information and hence the same connectivity information would be stored multiple times.

To optimize memory, we resort to a different approach that consists in combining the canonical form of the unlabeled topology with the sequence of labels. This approach introduces the problem of determining the order of labels because the canonical order of unlabeled topologies is not sufficient. To clarify this point, let us consider the two subgraphs in Fig. 3.2. Numbers represent the canonical order of vertices, while letters represent labels. Note that the order between 2 and 3 is ambiguous (1-3-2 would be a valid order as well) since by exchanging them we obtain the same unlabeled canonical form. The two labeled topologies are clearly isomorphic. However, the label sequences in the canonical orders are different (ABC vs. ACB).

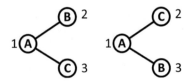

Fig. 3.2 Example of two unlabeled canonical orders that produce different sequence of labels on isomorphic graphs. The order is given by numbers. The two corresponding sequences of labels are ABC and ACB

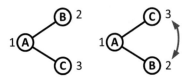

Fig. 3.3 By considering lexically ordered canonical orders we can guarantee that isomorphic graphs are associated with the same sequence of labels. In this example, both sequences of labels are ABC

To guarantee that isomorphic labeled topologies have the same label sequence, we solve the ambiguity in the canonical ordering by ordering labels (e.g., in alphabetic order) and using this order to break ties. This is equivalent to choosing, among all possible canonical orders (of the single canonical form) of an unlabeled topology, the one that corresponds to the lexicographically minimum sequence of labels. We refer to a canonical order that satisfies this condition as a *lexically ordered canonical order*. Note that network-centric tools (e.g., gTrie) solve the ambiguity by considering the order of vertex ids to break the ties. Therefore, we just need to ensure that the order of vertex ids is consistent with the order of labels. This can be done by reassigning vertex ids of the input network so that vertices with smaller labels are assigned with smaller vertex ids (i.e., $v \leq u$ if $l_{\mathbb{G}}(v) \leq l_{\mathbb{G}}(u)$). We call a graph that satisfies this condition a *lexically numbered graph*.

The procedure described above solves the problem in Fig. 3.2. The order of the second topology is forced to be as in Fig. 3.3 and hence both label sequences would be ABC. Now we prove that this procedure always gives the correct result. Specifically, we prove that

- if two labeled topologies are isomorphic, then their associated labeled canonical forms (topology + label sequence) are equal;
- given two labeled topologies, if their corresponding labeled canonical forms are equal, then they are isomorphic (including their labels).

The second condition is trivial. Indeed the canonical order of two topologies defines an association between vertices that preserves both the structure and the label sequence.

In order to prove the first condition, we need to prove that if two labeled topologies are isomorphic then their corresponding sequence of labels coincide. In fact, the labeled canonical form is computed by combining the unlabeled canonical form with the sequence of labels. Since by stripping off the labels two isomorphic topologies remain isomorphic, the two unlabeled canonical forms coincide. Therefore we need to be concerned only about the label sequences.

Lemma 3.1 *Let G_1, G_2 be two labeled subgraphs of a lexically numbered graph and S_1, S_2 be the sequence of labels given by their lexically ordered canonical order. If G_1 and G_2 are isomorphic, then $S_1 = S_2$.*

Proof By contradiction. Suppose $S_1 \neq S_2$. Without loss of generality, consider $S_1 < S_2$. Since G_1 and G_2 are isomorphic, there is at least an isomorphism between G_1 and G_2 (that is a one-to-one association between vertices of G_1 and vertices of G_2). We can use this isomorphism to construct an order of vertices for G_2 that is equivalent to the canonical order of G_1. This is a valid canonical order for G_2 since it produces the same unlabeled canonical form. Moreover this order corresponds to the same sequence of labels as S_1. That constitutes a valid canonical order for G_2 that produces a sequence of labels smaller than S_2. This contradicts the hypothesis that S_2 was obtained by a lexically ordered canonical order.   □

We modify the G-Trie algorithm to support labels. The main change concerns the information associated with gTrie nodes. Specifically, we substitute the counters of gTrie nodes with hash tables that associate label sequences to counters. To retrieve the counter of a labeled topology, we first look up the entry corresponding to its unlabeled topology, then we look up the counter associated with the label sequence in the corresponding hash table.

In summary, we apply the following changes to G-Trie:

1. Introduce a first step that reassigns ids to vertices of the input network so that vertices with smaller labels are assigned with smaller vertex ids (to create lexically numbered graphs).
2. Substitute the counters of gTrie nodes with hash tables that associate label sequences to counters.
3. Change the census procedure to increase the counters of labeled topologies as opposed as unlabeled ones.

Since the major changes are in the census algorithm we focus on the census procedure for labeled motif discovery (the overall algorithm for the labeled case is quite similar to that of the labeled case, but the labeled case requires the addition of the initial vertex ids assignment step). The census algorithm is shown in Algorithm 3.

Algorithm 3: Census algorithm for labeled motif discovery
***

Require: labeled network, gTrie {returns the gTrie filled with hash tables with the number
of occurrences of each labeled topology.}
$Match(gTrie.root, \emptyset)$
return $gTrie$

Procedure $Match(node, V_{used})$
if $V_{used} = \emptyset$ **then**
    $V_{cand} \leftarrow V(network)$
else
    $V_{cand} \leftarrow \{v \in N(V_{used}) : v$ satisfies symmetry breaking conditions$\}$
end if
$V \leftarrow \emptyset$
for all $v \in V_{cand}$ **do**
    if v is connected with V_{used} as defined in $node$ **then**
        $V \leftarrow V \cup \{v\}$
    end if
end for
for all $v \in V$ **do**
    if $isLeaf(node)$ **then**
        $label\_seq \leftarrow$ labels of V_{used} in lexically ordered canonical order
        $node.hash\_table[label\_seq]+ = 1$
        {now we have a hash table as opposed as a counter}
    end if
    for all children c of $node$ **do**
        $Match(c, V_{used} \cup \{v\})$
    end for
end for
End Procedure
***

3.2.5 An Index for Querying Motifs

During the preprocessing phase, we find all motifs up to size K (a pre-defined
parameter) in the input network as described previously. We set neither a frequency
threshold nor a p-value threshold at this point so that queriers can set thresholds of
interest at query time. One implication is that all labeled topologies occurring in the
input network having size K or less are considered. For simplicity of exposition,
in the following, we consider only motifs of size exactly K, although our method
handles motifs with size smaller than K, as we explain later. We put all extracted
labeled topologies in a data structure, which we call the *TopoIndex*, that facilitates
later retrieval. An example of a TopoIndex for $K = 3$ and two labels (A and B) is
depicted in Fig. 3.4.

The TopoIndex consists of a DAG, which embodies the super-multiset relation
between sets, and a collection of lists of topologies contained in the leaves of the
DAG. Specifically, nodes of the DAG represent bags of labels (label constraints)
and an edge is drawn between two nodes u and v if v is super-multiset of u (i.e., it
contains all labels in u with multiplicity below or attained to the one in v), and v has

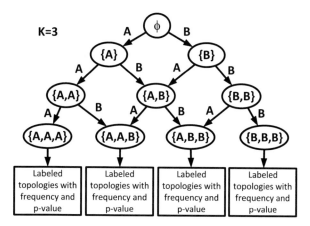

Fig. 3.4 The TopoIndex. Our data structure for processing label-based queries

exactly one label more than u. The edge is associated with the label that is different between u and v. Each leaf (node that does not have any outgoing edges) contains a list of all labeled topologies that satisfy the label constraints associated with the leaf, with the topologies' frequencies and p-values.

The described TopoIndex enables fast lookup of a bag of labels and then fast retrieval of associated topologies (by exploring the part of the DAG reachable from the corresponding node). The DAG shown in the example in Fig. 3.4 is complete, that is, it contains all possible nodes up to depth 3, but in general it may be not need to be complete. For instance, if there are no topologies with labels ABB and BBB, the nodes ABB, BBB, and BB are not included in the DAG, thus saving time and space.

3.2.5.1 Building the TopoIndex

The building procedure is given in Algorithm 4. First we group the topologies by their label bags. Then, for each label bag we create a leaf and store it in a hash table that associates label bags with the corresponding nodes. We create the other nodes of the DAG by calling *create_dag*() (Algorithm 5), which recursively removes one label at a time from nodes and creates nodes up to the root. The time complexity of Algorithm 4 is $O(|T||K|\log(|K|) + |LB||K|^2)$, where LB is the set of label bags ($|LB| \leq |T|$). The labels of every topology need to be ordered (for comparison with other label bags), which can be done in time complexity $O(|K|\log(|K|))$. Grouping label bags can be done in time $O(|T||K|)$ in expected time using a hash table. Inserting the label bags in the TopoIndex can be done in time $O(|LB||K|^2)$ since for every unlabeled topologies at most $|K|$ nodes need to be looked up by the recursive call *create_dag*() and looking up a node can be done in time $O(|K|)$. Since $|K|$ is usually very small, the building time is effectively linear over $|T|$.

Algorithm 4: Building the TopoIndex

Require: set T of labeled topologies of size K with associated frequency and p-value
{returns the root of the TopoIndex data structure}
group T by label bags
for each label bag lb and its corresponding set of topologies T_{lb} **do**
    initialize *node* {create a leaf node}
    node.label_bag = lb
    node.topologies = T_{lb}
    hash_table[lb] = *node*
    call *create_dag(node, hash_table)*
end for
return *hash_table*[{∅}]

Algorithm 5: Recursive procedure *create_dag* for building the TopoIndex

---

Require: a node *node* and the hash table of nodes *hash_table*
if *node.label_bag* == ∅ **then**
    return
end if
for each label l in *node.label_bag* **do**
    lb_parent = *node.label_bag* − {l}
    if *lb_parent* ∈ *keys(hash_table)* **then**
        parent = *hash_table*[*lb_parent*]
    else
        initialize *parent* {create a new node}
        parent.label_bag = *lb_parent*
        hash_table[*lb_parent*] = *parent*
        call *create_dag(parent, hash_table)*
    end if
    parent.children[l] = *node*
end for

3.2.5.2 Query Processing

Given the TopoIndex described above, and a query $Q = (C, k, f, p)$ with $k = K$, query processing is quite straightforward. To perform a query $Q = (C, k, f, p)$ with $k = K$, first look up the node n of the DAG associated with the set of labels in C, then explore all nodes of the DAG reachable from n. Finally, retrieve all topologies associated with reachable leaves and return the ones whose frequencies are greater than or equal to f and whose p-values are less than or equal to p.

The TopoIndex can be changed to support queries of size $k \leq K$ by associating internal nodes at depth k to labeled topologies of size k (for all $k = 1 \ldots K - 1$). Answering queries with $k > K$ is the subject of our current work.

3.3   Alternative Methods of Calculating Statistical Significance

One might ask why we care about statistical significance (reflected in the p-value calculation in the previous section). Studies have shown that in many biological networks, small subnetworks of real networks that are much more frequent than random networks of the same size (Alon 2007; Milo et al. 2002) often act as functionally important modules. For example, in Alon (2007) and Milo et al. (2002) the authors identified motifs representing positive and negative autoregulation (subnetworks of one node and one edge), coherent and incoherent feed-forward loops (subnetworks of three nodes and three edges), single-input modules (one node connected to few or many other nodes), and dense overlapping regulons (many nodes connected to few or many other nodes). One function of a coherent feed-forward loop formed by a target Z and two transcription factors X and Y is the logic operation AND of a circuit: Z is activated by both X and Y; however, Y is also regulated by X. Motif functionality has also been investigated with respect to evolution (Kashtan and Alon 2005; Solé et al. 2002) showing that motifs with the same topologies can have important functionality in different conditions.

That explains our interest in finding statistically overrepresented substructures. This section discusses approaches to establishing statistical significance.

Formally, given a graph $G = (V, E)$ (directed or undirected) with n vertices whose ids are uniquely labeled with integers from 1 to n. A connected subgraph induced by a set of vertices of cardinality k (a *topology* for short) is called a motif when it occurs statistically significantly more often than the same subgraph in randomized networks derived from the original network (Milo et al. 2003).

The random generation method to find motifs given a real network consists of the following steps: (1) generate a large set of random networks that share the characteristics of the real network; (2) find candidate topologies, consisting of subgraphs in the real network; (3) count the occurrences of these topologies; (4) assess the significance of each topology by computing its number of occurrences in each of the random networks.

The first step creates networks that have the same number of nodes and edges of the real network. Moreover, each node in the generated network maintains its original number of edges leaving and entering the node (Newman et al. 2001). Next, by proceeding in an exhaustive manner, an algorithm can define all possible topologies of subgraphs with n nodes and count all the occurrences of such subgraphs in the real and in the random networks (Milo et al. 2003).

The random generation method consists of two expensive steps: the generation of a large number of networks and the application of subgraph isomorphism algorithms to compute the number of occurrences. Over the last decades, researchers have worked to reduce the expense of both steps. We list the main results in the following sections. For the sake of brevity, we point to the main alternative approaches, but give few details.

3.3.1 Quasi-Analytical Methods to Assess the Statistical Significance of a Topology

The random generation method described above evaluates the significance of the topology through the computation of a z-score using a Gaussian assumption or a p-value using a resampling approach (Milo et al. 2002, 2003; Prill et al. 2005; Shen-Orr et al. 2002). The Gaussian assumption may not apply to a particular application, but a reliable p-value requires a large number of random graphs whose analysis turns out to be computational expensive (by far more expensive than analyzing the target network alone). Recently, researchers have investigated the possibility of analyzing the distribution of the topologies, both noninduced and induced, from an analytical point of view that would avoid the need for random generation. Table 3.1 summarizes the main ideas of the two above approaches.

Approximation methods, based on the Erdos–Renyi (ED) model, have studied the asymptotic normality of the distribution of the count of the topologies (Wernicke 2006). Unfortunately, the Erdos–Renyi random model is a poor approximation to some networks of interest, such as biological networks (Barabási and Albert 1999). Alternative reference models include the fixed degree Distribution (FDD) (Newman et al. 2001) that models the random generation method of swapping random edges. The swapping approach guarantees that a given node has the same valence in the random graphs as in the original one. There is also a variant of the FDD called Expected Degree Distribution (EDD) (Picard et al. 2008) and the Erdos–Renyi Mixture for Graphs (ERMG) (Picard et al. 2008). Table 3.2 depicts the main features and differences of the models.

The EDD model generates random graphs whose degrees follow the distribution of the original graph, but particular nodes may obtain different valences. Conditional to the distribution of node degrees, the probability of edges is modeled as independent and exists with a probability proportional to the product of the degree distributions of the involved nodes. In the ERMG model, the nodes are spread among Q hidden classes with respective proportions $\alpha_1, \cdots, \alpha_Q$. The edges are independent conditional on the class of the nodes. The connection probability depends on the classes of both nodes.

Table 3.1 P-value generation

	Sampling + Permutation test	Analytical model
Idea	Generate random graph according to some random model. P-value is the fraction of graphs in which the occurrences in the random graphs is higher than the target one	The target graph belongs to a given distribution. Define a Random Variable representing the number of occurrences of the motif under the reference model
Pros	Easy to implement	Computationally inexpensive
Cons	Computationally expensive	May not be possible to identify an appropriate distribution

Simulation vs analytics approach

Table 3.2 Random Models: ER=Erdos–Renyi, FDD-Fixed Degree Distribution, EDD=Expected Degree Distribution, EMGR=Erdos–Renyi Mixture for Graphs

Name	Characteristics	Graph Distribution generation
ER	All edges of a graph are independent and exist with probability p.	The connection probability p of the ER model is estimated by the proportion of observed edges in the network.
FDD	Generates graphs whose degrees have exactly a given distribution.	For the given sequence of degrees in the input network, the graph is chosen uniformly at random from the set of all graphs with that degree sequence.
EDD	Generates graphs whose degrees follow a given distribution.	The empirical distribution of the degrees in the network is used as the distribution of the expected degrees.
EMGR	Nodes are spread among Q hidden classes with respective proportions $p\_1, \ldots, p\_Q$. Edges are independent conditionally to the class of the nodes. The connection probability depends on the classes of both nodes.	Fit a mixture model.

It has been also shown that the use of the Compound-Poisson distribution (Adelson 1966) in the Erdos–Renyi random model allows the accurate approximation of the number of overrepresented topologies (Picard et al. 2008). In Picard et al. (2008), the authors propose a model for the exact calculation of the mean and variance under any model of exchangeable random graphs (exchangeability means that the probability of occurrence of a topology does not dependent on its position in the graph, i.e., on the topological structure of the neighborhood of the topology). Furthermore, the authors have shown that the Polya–Aeppli distribution (also known as the Poisson Geometric distribution, which is a special case of the Poisson-Compound distribution) is a good model for the distribution of the count of the topologies (both induced and noninduced) and leads to a more accurate p-value than a Gaussian model for the graphs of many applications. The reason is that the Geometric-Poisson distribution is particularly suitable for describing the number of events that occur in clusters, where a Poisson distribution describes the number of clusters and the counts of events within a cluster follow a geometric distribution. Here, this fits the case when distinct topologies can share nodes and edges (i.e. clumps) (Picard et al. 2008). In fact, the authors show that when the number of clumps has a Poisson distribution with mean λ and the sizes of the clumps are independent of each other and have a Geometric distribution $G(1 - a)$, the number of observed events X (topologies) has a distribution $P(\lambda, a)$ and leads to an estimate of the number of occurrences of a given topology (see Table 3.1).

So far, our discussion has concerned label-free (also known as color-free) networks. Schbath et al. (2009) propose an analytical model for the computation of p-values for colored patterns. A colored pattern is a topology having a given multiset of colors (vertex labels). For example, a star of size 5 having 4 Bs and 1 C. An occurrence of the pattern is defined as a connected subgraph whose labels have a match with the multiset. Schbath et al. would make no distinction between a star having the C in the center or one having the C on the outside. That subtle difference makes our job more difficult, but the starting point for our current research is their excellent work.

Schabat et al. define analytical formulas for the mean and variance of the number of colored topologies by using the Erdos–Renyi model. Thanks to this, they were able to derive a reliable z-score for each topology. The authors then model the distribution of the count of colored topologies under the Erdos–Renyi model.

3.3.2   Random Generation Methods

Whereas the previous subsection discussed analytical method, no published analytical method can discover p-values under our model of query (though, as mentioned, we ourselves are making progress toward that goal). So we turn to random generation methods. Improving random generation methods entails intelligent searching through graphs to enumerate topologies. The basic idea is to start from single nodes and expand them with their neighborhoods in a tree-like fashion, checking at each step that each subgraph in the tree appears only once and that it does not violate the color constraints of the query. This procedure can be further improved by sampling the network (Alon 2007) or the neighborhoods in the expanding phases (Wernicke 2006). Alternatively, Grochow and Kellis (2007) used subgraph enumeration and symmetry breaking to avoid the search for automorphisms of the subgraphs occurrences. We now give some examples of the state-of-the-art algorithms upon which we build our structure.

The ESU algorithm (Wernicke 2006) enumerates all subgraphs of size k by starting from a root vertex v of the graph and computing the occurrences of the topology by extending it node by node. The algorithm uses the concept of exclusive neighborhood, which is defined as follows. For a subset $V' \subseteq V$, its open neighborhood $N(V')$ is the set of vertices in $V \setminus V'$, which are adjacent to at least one vertex in V'. For each node $v \in V \setminus V'$, the exclusive neighborhood with respect to V' and denoted by $N_{excl}(v, V')$ consists of all vertices that are neighbors of v but are not in $V \cup N(V')$ (Fig. 3.5).

The key idea of the algorithm is to add into the extension set of v, called $V_{Extension}$, only those vertices satisfying the two following properties: (1) their vertex ids must be greater than v; (2) must be neighbors only to the newly added w and not already in $V_{subgraph}$ (i.e. they must be in $N(w, V_{subgraph})$).

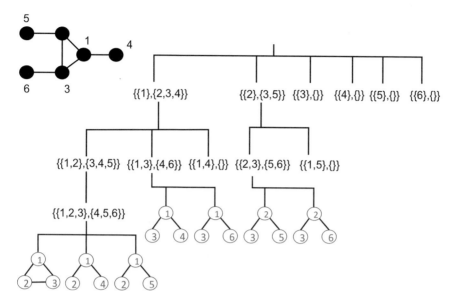

Fig. 3.5 The ESU tree for generating all subgraphs of $k=3$ nodes

Its randomized variant, Rand-ESU, introduces an option that performs a uniform sampling in the graph, thus avoiding the need to explore it all. The algorithm is essentially the same as the original one with the exception that the recursion is carried out with a certain probability that decreases with the depth of the enumeration. In practice, the probability is high in the first steps of the recursion and then decreases as the size of the subgraphs to be explored increases.

The sampling in RAND-ESU is unbiased and is quite simple to implement. On the other hand, RAND-ESU gives only an estimate of the number of occurrences.

Graph mining algorithms (Yan and Han 2002) find frequent subgraphs in a database of graphs or in a single large graph. A subgraph is frequent if its support (occurrence frequency) in a given dataset (or in a graph) is no less than a *minimum support* threshold. Computing the statistical significance of such topologies is done by simulation, as described above.

In this chapter, we consider the problem of searching for topologies of labeled graphs. However, there are several possible definitions of labeled topology.

In Schbath et al. (2009), the authors define a potential k-colored motif to be any connected subgraph of k nodes containing a specified multiset of colors (defined on the nodes). The motif is "potential" because its statistical significance may not meet a threshold. In this case, different topologies with the same labels define the same motif. Adami et al. (2011) consider the definition of colored motif as above, and use a measure based on entropy to determine the significance. In Wernicke (2006) and Ribeiro and Silva (2014a), the authors use the definition of motifs colored on both nodes and edges having a specific topology. Wernicke (2006) is based on the ESU algorithm, whereas Ribeiro and Silva (2014a) introduce a version of GTrie capable to find colored motifs.

In this chapter, we adopt the motif definition introduced in Ribeiro and Silva (2014a).

Definition 3.2 Let G be a labeled graph. Let $m(V_m, E_m, LV_m, LE_m)$ be a subgraph of G with V_m nodes and E_m edges, where LV_m and LE_m are two sets of colors representing the labels of nodes and edges, respectively. Let c be the number of isomorphic occurrences of m in G, and let α be a critical value. Let G_R be a random variant of G obtained by applying the edge shuffling method based on the Fixed Degree Distribution, and let c_R be the number of occurrences of m in the random variant G_R. We say that m is a motif of G if, by applying a permutation test using k random variant of G, $G_{R,i}$ ($k = 500$ usually), $\frac{\#(c_{R,i} > c)}{k} < \alpha$, where $\#(c_{R,i} > c)$ is the number of times the number of occurrences of m in $G_{R,i}$ is greater than in G.

Because there is no analytical way to compute the significance of such a network motif yet, we will use the simulation on the random generated networks to establish the significance of colored network topologies. Algorithm 6 shows the implementation of a permutation test.

In our current efforts, we extend the analytical approach of Schbath et al. (2009) and Picard et al. (2008) to compute the significance of topologies given a multiset of colors.

Algorithm 6: Randomized generation test to discover p-values

Require: network G, candidate topologies $m_1, m_2, \cdots, m_l$, c_i number of occurrences of m_i in G, number of iterations k, critical value α {returns the p-value of topology m_j}

$s_j := 0$

for $j = 1, \ldots l$ **do**

  for $i = 0 \ldots k$ **do**

    $G_{R,i} = randomize(G)$

    $c_{R,j} :=$ number of occurrences of m_j in $G_{R,i}$;

    for $j = 0 \ldots l$ **do**

      if $c_{R,j} \geq c_j$ **then**

        $s_j + +$

      end if

    end for

  end for

end for

for $j = 0 \ldots l$ **do**

  output p-value of topology m_j is s_j/k

end for

3.4 Experiments

gLabTrie has been tested on a dataset of social, communication, and biological networks. All experiments has been performed on the following configuration: Intel Core i7-2670 2.2 Ghz CPU with a RAM of 8 GB. Table 3.3 describes the features of the selected networks.

FLIGHTS is a network extracted from Openflights.org (http://openflights.org), representing all possible air routes between different airports around the world in 2011 (Opsahl 2011). BLOGS is a directed network of hyperlinks between web logs on US politics of 2004 (Adamic and Glance 2005). PPI is a protein–protein interaction (PPI) network in human, taken from HPRD database (Keshava Prasad et al. 2009). DBLP is the citation network of DBLP, a database of scientific publications, where each node in the network is a publication and edges connect two citations A and B iff A cites B (Ley 2002). FOLDOC is an oriented semantic network taken from the on-line computing dictionary FOLDOC (http://foldoc.org), where nodes are computer science terms and edges connect two terms X and Y iff Y is used to explain the meaning of X (Batagelj et al. 2002). INTERNET represents the business relationships between autonomous systems (ASes) of Internet in 2005 (Dimitropoulos et al. 2005).

Nodes of each network have been annotated with the following labels. In FLIGHTS, airports have been associated to one of the five continents. In BLOGS, nodes have been classified depending on their political leaning (liberal and conservative). For the labeling of nodes in PPI, we used Gene Ontology (GO) (Ashburner et al. 2000), a hierarchical dictionary of terms related to biological processes, components, and functions, which have been extensively used for the analysis of biological networks so far (Maere et al. 2005; Bindea et al. 2009). We annotated proteins with GO processes up to the first level of the hierarchy yielding 11 nodes labels. Ten of them represent specific kinds of biological processes (whole-organism process, metabolism, regulation, cellular organization, development, localization, signaling, response to stimulus, biological adhesion, and reproduction). A special label representing the generic biological process has been associated to proteins for which we did not have GO annotations. DBLP nodes has been annotated with different kinds of publications (articles, inproceedings, proceedings, books, incollections, PhD thesis, and master thesis) or "www" if the node refers to a cited

Table 3.3 Networks used for experiments

Name	Type	Nodes	Edges	Reference
FLIGHTS	Undirected	2939	15,677	Opsahl (2011)
BLOGS	Directed	1224	16,715	Adamic and Glance (2005)
PPI	Undirected	9506	37,054	Keshava Prasad et al. (2009)
DBLP	Directed	12,591	49,728	Ley (2002)
FOLDOC	Directed	13,356	120,239	Batagelj et al. (2002)
INTERNET	Undirected	20,305	42,568	Dimitropoulos et al. (2005)

web site. INTERNET ASes have been partitioned into seven classes (large ISPs, small ISPs, customers, universities, Internet exchange points, network information centers, not classified) according to the taxonomy described in Dimitropoulos et al. (2006). Computing terms in FOLDOC have been labeled according to their domains (jargons, computer science, hardware, programming, graphics and multimedia, science, people and organizations, data, networking, documentation, operating systems, languages, software, various terms).

We compared the no-index version of gLabTrie with the index-based approach. We run our algorithm using default randomization parameters ($N_{rand} = 100$, $p = 0.01$ and $f = 2$).

The performance of gLabTrie has been evaluated with respect to three parameters:

(a) m: the motif size, i.e., the number of its nodes
(b) c: the number of motif constraints, i.e., the number of specified node labels in the query
(c) l: the number of labels in the input networks

For tests (a) and (b) we used real labels, while in case (c) we ran our algorithm with randomly assigned labels. To measure the influence of these parameters, we varied the parameter of interest and assigned default values to the other ones ($m = 4$, $c = 4$ and $l = 2$). For each test, we ran gLabTrie on a set of 10 random queries. In the experiments with real labels, label constraints for random queries were generated according to the frequency of a node label: the more frequent a label x, the higher the probability that x is added as a label constraint to the query. In the tests with artificial labels, label constraints were added to the queries according to the uniform distribution of node labels.

Table 3.4 reports the running times for building indexes for motif of size m up to 4 in networks annotated with real labels. In all cases, the performance of gLabTrie strongly depends on the size of the network, its orientation (undirected graphs contain more instances of a certain topology on average), and the number of labels. Most of the time is spent in storing all the motif occurrences of a given size into the database. The number of occurrences increases exponentially with m.

Table 3.5 shows the results of the comparison between the no-index and the index-based approach of gLabTrie on querying motifs of different sizes, up to size 4. For each network and each motif size, we reported the mean and the standard

Table 3.4 Running times (minutes) to build indexes on varying motif size

Network	$m = 3$	$m = 4$
FLIGHTS	8.59	245.39
BLOGS	7.78	566.83
PPI	21.72	425.59
DBLP	30.91	1211.91
FOLDOC	46.28	1486.59
INTERNET	87.48	40,605.23

Table 3.5 Running times (s) for querying motifs of different size with no-index and index-based approach

Network	m	No-index	Index
FLIGHTS	3	333.02 ± 4.71	0.01 ± 0.01
	4	364.81 ± 38.70	0.56 ± 0.86
BLOGS	3	155.36 ± 2.27	0.08 ± 0.15
	4	960.07 ± 159.01	1.44 ± 0.54
PPI	3	872.68 ± 21.77	0.01 ± 0.01
	4	866.10 ± 5.17	0.06 ± 0.10
DBLP	3	553.46 ± 4.02	0.11 ± 0.09
	4	882.63 ± 152.05	6.28 ± 5.06
FOLDOC	3	1290.23 ± 8.45	0.02 ± 0.01
	4	1308.40 ± 12.55	0.75 ± 0.22
INTERNET	3	2116.65 ± 6.38	0.70 ± 2.04
	4	2649.60 ± 1305.87	670.22 ± 200.59

deviation. In both cases, the running time includes the time needed to retrieve all the subgraphs matching a given query.

The results show (unsurprisingly) that the index-based approach is much faster (100s of times) than having no index. We define q_{min} to be the minimum number of query operations required to recoup the time cost of building the index. For $m = 3$, $q_{min} \simeq 2$, so the time cost of building the index is recouped after two queries on average, while for $m = 4$ we have $q_{min} \simeq 44$.

It is worth noting that the benefit of the index decreases as the size of the network (measured in terms of the number of its nodes) increases. For instance, in the INTERNET network, which is by far the biggest network in our dataset, when $m = 4$ the index-based approach is only four times faster than the no-index one. In this case, the disappointing performance of the index-based approach is due to the very high number of query occurrences that the algorithm must retrieve from the dataset, resulting in a large number of I/O operations. In the INTERNET network with $m = 4$ the I/O time is 99% of the total running time, on average.

In Table 3.6, we compare the running times of the no-index and the index-based approach on querying motifs with a variable number of label constraints in the query. Again, network nodes have been annotated with real labels. We set $m = 4$ and we varied c from 1 to 4.

As the number of query label constraints defined by the user increases, the performance of both approaches improves. However, the more selective the query, the greater is the benefit of the index. The gain enjoyed by the index is proportional to the size of the network and the number of constraints, because of the exponential decrease of the number of occurrences matching the query. For example, when c goes from 1 to 4, the no-index approach becomes $\simeq 28$ times faster and the index-based approach $\simeq 16400$ times faster in the INTERNET network, while in the BLOGS network the two algorithms are only $\simeq 3$ and $\simeq 15$ faster, respectively.

Table 3.6 Running times (s) for querying motifs with variable number of label constraints with no-index and index-based approach

Network	c	No-index	Index
FLIGHTS	1	685.06 ± 155.22	13.35 ± 9.6
	2	576.76 ± 116.71	5.70 ± 4.93
	3	412.65 ± 61.02	2.09 ± 3.32
	4	343.85 ± 17.17	0.52 ± 1.06
BLOGS	1	2214.74 ± 8.36	48.32 ± 2.75
	2	1953.47 ± 199.11	21.20 ± 20.42
	3	1430.81 ± 336.02	15.83 ± 20.09
	4	829.10 ± 214.59	6.16 ± 13.74
PPI	1	1228.73 ± 221.05	10.79 ± 10.18
	2	1116.63 ± 216.56	9.35 ± 11.52
	3	897.56 ± 34.87	0.51 ± 0.75
	4	861.30 ± 9.43	0.04 ± 0.06
DBLP	1	4041.38 ± 316.75	139.08 ± 1.84
	2	3186.24 ± 693.96	80.60 ± 34.95
	3	1867.96 ± 475.73	40.26 ± 18.37
	4	871.19 ± 131.45	7.43 ± 4.14
FOLDOC	1	3751.95 ± 686.84	78.29 ± 50.07
	2	2075.93 ± 334.84	7.24 ± 2.42
	3	1288.12 ± 42.41	2.47 ± 2.15
	4	1212.23 ± 16.79	0.58 ± 0.46
INTERNET	1	57,988.44 ± 13,722.07	11,642.50 ± 5519.74
	2	28,082.31 ± 13,974.45	8984.35 ± 6945.80
	3	9165.14 ± 5849.21	3660.06 ± 3297.24
	4	2101.29 ± 35.61	0.71 ± 1.05

Table 3.7 summarizes the results of the comparison between the performance of the two approaches when the number of labels vary. To perform these experiments, we annotated network nodes with artificial labels. Given a set of l labels, each node has been associated with a random unique label between 1 and l, according to a uniform distribution. We ran five different series of experiments with $l = 2, 6, 10, 14, 18$. In each series, we set $m = 4$ and $c = 4$.

The time costs of both approaches decrease when the number of node labels increase. In all networks, the greatest reduction of the running time happens when we move from $l = 2$ to $l = 6$.

Table 3.7 Running times (s) for querying motifs with variable number of node labels with no-index and index-based approach

Network	l	No-index	Index
FLIGHTS	2	483.08 ± 35.09	4.97 ± 1.29
	6	331.79 ± 3.18	0.09 ± 0.03
	10	327.36 ± 0.63	0.03 ± 0.01
	14	326.89 ± 0.39	0.04 ± 0.02
	18	327.00 ± 0.82	0.09 ± 0.02
BLOGS	2	931.77 ± 254.88	7.37 ± 2.32
	6	192.27 ± 29.18	0.29 ± 0.18
	10	160.67 ± 5.44	0.55 ± 0.05
	14	151.54 ± 2.77	1.16 ± 0.12
	18	149.44 ± 1.36	2.25 ± 0.10
PPI	2	1066.70 ± 95.43	2.10 ± 1.66
	6	870.64 ± 5.42	0.17 ± 0.12
	10	861.66 ± 1.72	0.04 ± 0.02
	14	860.80 ± 1.55	0.05 ± 0.03
	18	851.23 ± 2.16	0.10 ± 0.07
DBLP	2	1686.55 ± 496.97	32.66 ± 20.59
	6	623.22 ± 25.33	1.59 ± 0.94
	10	571.75 ± 11.92	1.23 ± 0.76
	14	559.46 ± 5.47	0.57 ± 0.36
	18	555.38 ± 2.36	1.32 ± 0.64
FOLDOC	2	2749.55 ± 539.16	18.67 ± 6.22
	6	1266.62 ± 43.13	0.81 ± 0.83
	10	1218.74 ± 10.16	1.01 ± 0.80
	14	1204.28 ± 6.85	1.63 ± 0.88
	18	1201.15 ± 1.64	3.07 ± 0.98
INTERNET	2	17270.40 ± 5731.44	1154.77 ± 2030.07
	6	2595.38 ± 382.33	94.18 ± 106.23
	10	2250.07 ± 111.54	23.98 ± 20.00
	14	2162.81 ± 44.00	5.68 ± 4.18
	18	2113.16 ± 20.06	5.08 ± 4.59

3.5   Conclusion

Our structures gLabTrie and TopoIndex contribute to all aspects of motif finding, by giving a very fast method for finding labeled topological structures in both input networks and related random networks. As this is work in progress, we plan in the near future to (1) find analytical methods for computing p-values on labeled topological structures to avoid the need for random graphs; (2) extend the search algorithms to enable search for topologies having, say, k vertices, even though the TopoIndex holds topologies of only a smaller size.

Acknowledgements Shasha's work has been partially supported by an INRIA International Chair and the U.S. National Science Foundation under grants MCB-1412232, IOS-1339362, MCB-1355462, MCB-1158273, IOS-0922738, and MCB-0929339. This support is greatly appreciated.

References

Adami C, Qian J, Rupp M, Hintze A (2011) Information content of colored motifs in complex networks. Artif Life 17(4):375–390

Adamic LA, Glance N (2005) The political blogosphere and the 2004 u.s. election: divided they blog. In: Proceedings of the 3rd international workshop on link discovery, LinkKDD '05. ACM, New York, pp 36–43

Adelson RM (1966) Compound Poisson distributions. Oper Res Q 17(1):73–75

Alon U (2007) Network motifs: theory and experimental approaches. Nat Rev Genet 8(6):450–461

Ashburner M, Ball C, Blake J, Botstein D, Butler H, Cherry J, Davis A, Dolinski K, Dwight S, Eppig J, Harris M, Hill D, Issel-Tarver L, Kasarskis A, Lewis S, Matese J, Richardson J, Ringwald M, Rubin G, Sherlock G (2000) Gene ontology: tool for the unification of biology. Nat Genet 25(1):25–29

Barabási AL, Albert R (1999) Emergence of scaling in random networks. Science 286(5439):509–512

Batagelj V, Mrvar A, Zaversnik M (2002) Network analysis of dictionaries. In: Language technologies, pp 135–142

Bindea G, Mlecnik B, Hackl H, Charoentong P, Tosolini M, Kirilovsky A, Fridman WH, Pages F, Trajanoski Z, Galon J (2009) ClueGO: a cytoscape plug-in to decipher functionally grouped gene ontology and pathway annotation networks. Bioinformatics 25(8):1091–1093

Dimitropoulos X, Krioukov D, Huffaker B, Claffy K, Riley G (2005) Inferring AS relationships: dead end or lively beginning? In: Nikoletseas SE (ed) Experimental and efficient algorithms. Springer, Berlin, pp 113–125

Dimitropoulos XA, Krioukov DV, Riley GF, Claffy KC (2006) Revealing the autonomous system taxonomy: the machine learning approach. CoRR abs/cs/0604015

Grochow JA, Kellis M (2007) Network motif discovery using subgraph enumeration and symmetry-breaking. In: Speed T, Huang H (eds) Research in computational molecular biology. Springer, Berlin, pp 92–106

Kashtan N, Alon U (2005) Spontaneous evolution of modularity and network motifs. Proc Natl Acad Sci 102(39):13773–13778

Keshava Prasad TS, Goel R, Kandasamy K, Keerthikumar S, Kumar S, Mathivanan S, Telikicherla D, Raju R, Shafreen B, Venugopal A, Balakrishnan L, Marimuthu A, Banerjee S, Somanathan DS, Sebastian A, Rani S, Ray S, Harrys Kishore CJ, Kanth S, Ahmed M, Kashyap MK, Mohmood R, Ramachandra YL, Krishna V, Rahiman BA, Mohan S, Ranganathan P, Ramabadran S, Chaerkady R, Pandey A (2009) Human protein reference database–2009 update. Nucleic Acids Res 37(Database issue):D767–772

Kurata H, Maeda K, Onaka T, Takata T (2014) BioFNet: biological functional network database for analysis and synthesis of biological systems. Brief Bioinform 15(5):699–709

Ley M (2002) The DBLP computer science bibliography: evolution, research issues, perspectives. In: Laender AHF, Oliveira AL (eds) String processing and information retrieval. Springer, Berlin, pp 1–10

Maere S, Heymans K, Kuiper M (2005) BiNGO: a cytoscape plugin to assess overrepresentation of gene ontology categories in biological networks. Bioinformatics 21(16):3448–3449

McKay BD (1981) Practical graph isomorphism. Congressus numerantium 30:45–87

Milo R, Shen-Orr S, Itzkovitz S, Kashtan N, Chklovskii D, Alon U (2002) Network motifs: simple building blocks of complex networks. Science 298(5594):824–827

Milo R, Kashtan N, Itzkovitz S, Newman MEJ, Alon U (2003) On the uniform generation of random graphs with prescribed degree sequences. eprint arXiv:cond-mat/0312028

Newman MEJ, Strogatz SH, Watts DJ (2001) Random graphs with arbitrary degree distributions and their applications. Phys Rev E 64:026118

Opsahl T (2011) Why anchorage is not (that) important: binary ties and sample selection. https://toreopsahl.com/2011/08/12/

Picard F, Daudin JJ, Koskas M, Schbath S, Robin S (2008) Assessing the exceptionality of network motifs. J Comput Biol 15(1):1–20

Prill RJ, Iglesias PA, Levchenko A (2005) Dynamic properties of network motifs contribute to biological network organization. PLOS Biol 3(11):e343

Ribeiro P, Silva F (2012) Querying subgraph sets with g-tries. In: Proceedings of the 2Nd ACM SIGMOD workshop on databases and social networks, DBSocial '12. ACM, New York, pp 25–30

Ribeiro P, Silva F (2014a) Discovering colored network motifs. In: Contucci P, Menezes R, Omicini A, Poncela-Casasnovas J (eds) Complex networks V. Springer International Publishing, Cham, pp 107–118

Ribeiro P, Silva F (2014b) G-Tries: a data structure for storing and finding subgraphs. Data Min Knowl Discov 28(2):337–377

Schbath S, Lacroix V, Sagot MF (2009) Assessing the exceptionality of coloured motifs in networks. EURASIP J Bioinform Syst Biol 2009:3:1–3:9

Shen-Orr SS, Milo R, Mangan S, Alon U (2002) Network motifs in the transcriptional regulation network of Escherichia coli. Nat Genet 31(1):64–68

Solé RV, Pastor-Satorras R, Smith E, Kepler TB (2002) A model of large-scale proteome evolution. Adv Complex Syst 05(01):43–54

Wernicke S (2006) Efficient detection of network motifs. IEEE/ACM Trans Comput Biol Bioinform 3(4):347–359

Yan X, Han J (2002) gSpan: graph-based substructure pattern mining. In: Proceedings - 2002 IEEE international conference on data mining. ICDM 2002, pp 721–724

Chapter 4
Applications of Flexible Querying to Graph Data

Alexandra Poulovassilis

Abstract Graph data models provide flexibility and extensibility, which makes them well-suited to modelling data that may be irregular, complex, and evolving in structure and content. However, a consequence of this is that users may not be familiar with the full structure of the data, which itself may be changing over time, making it hard for users to formulate queries that precisely match the data graph and meet their information-seeking requirements. There is a need, therefore, for flexible querying systems over graph data that can automatically make changes to the user's query so as to find additional or different answers, and so help the user to retrieve information of relevance to them. This chapter describes recent work in this area, looking at a variety of graph query languages, applications, flexible querying techniques and implementations.

4.1 Introduction

Due to their fine modelling granularity (in its simplest form, comprising just nodes and edges, naturally representing entities and relationships), graph data models provide flexibility and extensibility, which makes them well-suited for modelling complex, dynamically evolving datasets. Moreover, graph data models are typically semi-structured: there may not be a schema associated with the data; if there is a schema, then aspects of it may be missing from parts of the data and, conversely, parts of the data may not correspond to the schema. This makes graph data models well-suited to modelling heterogeneous and irregular datasets. Graph data models place a greater focus on the relationships between entities than other approaches to data modelling, viewing relationships as important as the entities themselves.

In recent years there has been a resurgence of academic and industry interest in graph databases, due to the generation of large volumes of data from web-based,

A. Poulovassilis (✉)
Birkbeck, University of London, London, UK
e-mail: ap@dcs.bbk.ac.uk

© Springer International Publishing AG, part of Springer Nature 2018
G. Fletcher et al. (eds.), *Graph Data Management*, Data-Centric Systems
and Applications, https://doi.org/10.1007/978-3-319-96193-4_4

mobile and pervasive applications centred on the relationships between entities, for example: the web graph itself; RDF Linked Data[1]; social and collaboration networks[2] (Martin et al. 2011; Suthers 2015); transportation and communication networks (Deo 2004); biological networks (Lacroix et al. 2004; Leser and Trissl 2009); workflows and business processes (Vanhatalo et al. 2008); customer relationship networks (Wu et al. 2009); intelligence networks (Ayers 1997; Chen et al. 2011); and much more![3]

As the volume of graph-structured data continues to grow, users may not be aware of its full details and may need to be assisted by querying systems which do not require queries to match exactly the data structures being queried, but rather can automatically make changes to the query so as to help the user find the information being sought. The OPTIONAL clause of SPARQL (Harris and Seaborne 2013) has the aim of returning matchings to a query that may fail to match some of the query's triple patterns. However, it is possible to "relax" a SPARQL query in ways other than just ignoring optional triple patterns, for example, making use of the knowledge encoded in an ontology associated with the data in order to replace an occurrence of a class in the query by a superclass, or an occurrence of a property by a superproperty.

This observation motivated the introduction in Hurtado et al. (2008) of a RELAX clause for querying RDF data, which can be applied to those triple patterns of a query that the user would like to be matched flexibly. These triple patterns are successively made more general so that the overall query returns successively more answers, at increasing 'costs' from the exact form of the query. We review this work on ontology-based query relaxation in this section, starting with an example application in heterogeneous data integration in Sect. 4.1.1.

Section 4.2 goes beyond conjunctive queries to consider conjunctive *regular path queries* over graph data, and *approximate answering* of such queries. In contrast to query relaxation, which generally returns additional answers compared to the exact form of a database query, query approximation returns potentially *different* answers to the exact form of a query.

Section 4.3 considers combining both query relaxation and approximate answering for conjunctive regular path queries over graph data, describing also an automaton-based implementation. Section 4.4 considers extending SPARQL 1.1 with query relaxation and approximation, describing an implementation based on query rewriting. Along the way, we consider applications of query relaxation and query approximation for graph data in areas such as heterogeneous data integration, ontology querying, educational networks, transport networks and analysis of user–system interactions. Section 4.5 covers additional topics: possible user interfaces for supporting users in incrementally constructing and understanding flexible queries

---

[1] http://linkeddata.org, http://www.w3.org/standards/semanticweb, accessed at 18/6/2015.

[2] https://snap.stanford.edu/data, accessed at 18/6/2015.

[3] See for example http://neo4j.com/use-cases, http://www.objectivity.com/products/infinitegraph, http://allegrograph.com/allegrograph-at-work, accessed at 18/6/2015.

and the answers being returned; and possible extensions to the query languages considered so far with additional flexibility beyond relaxation and approximation, and with additional expressivity in the form of path variables. Section 4.6 gives an overview of related work on query languages for graph data and flexible querying of such data. Section 4.7 gives our concluding remarks and possible directions of future work.

Flexible Database Query Processing

Before beginning our discussion of flexible query processing for graph data, we first review the main approaches to flexible query processing for other kinds of data. Due to the considerable breadth of this area, the references cited here are representative of the approaches discussed rather than an exhaustive list. Readers are referred to the proceedings of the bi-annual conference on Flexible Query Answering Systems (FQAS) for a broad coverage of work in this area.

Query languages for structured data models, such as SQL and OQL, include WHERE clauses that allow filtering criteria to be applied to the data matched by their SELECT clauses. Therefore, a natural way to relax queries expressed in such languages is by dropping a selection criterion, or by 'widening' a selection criterion so as to match a broader range of values (Bosc and Pivert 1992; Heer et al. 2008). Another common approach to query relaxation is to allow *fuzzy* matching of selection criteria, accompanied by a scoring function that determines the degree of matching of the returned query answers (Galindo et al. 1998; Na and Park 2005; Bordogna and Psaila 2008; Bosc et al. 2009). Conversely, queries can be made more specific by adding user preferences as additional filter conditions, with possibly fuzzy matching of such conditions (Mishra and Koudas 2009; Eckhardt et al. 2011). Chu et al. (1996) use type abstraction hierarchies to both generalise and specialise queries, while Zhou et al. (2007) explore statistically based query relaxation through 'malleable' schemas containing overlapping definitions of data structures and attributes.

Turning to approximate query answering, approaches include histograms (Ioannidis and Poosala 1999), wavelets (Chakrabarti et al. 2001) and sampling (Babcock et al. 2003). Sassi et al. (2012) describe a system that enables the user to issue an SQL aggregation query, see results as they are being produced, and dynamically control query execution. Fink and Olteanu (2011) study approximation of conjunctive queries on probabilistic databases by specifying lower- and upper-bound queries that can be computed more efficiently.

In principle, techniques proposed for flexible querying of structured data can also be applied to graph-structured data. However, such techniques do not focus on the connections (i.e. edges and paths) inherent in graph-structured data, thus missing opportunities for further supporting the user through approximation or relaxation of the path structure that may be present in a graph query.

Semi-structured data models aim to support data that are self-describing and that need not rigidly conform to a schema (Abiteboul et al. 1997; Buneman et al. 2000; Fernandez et al. 2000; Bray et al. 2008). Generally, such data can be modelled as a tree, though cyclic connections between nodes may also be allowed by the model (e.g. in XML, through the ID/IDREF constructs). Much work has been done on relaxing tree-pattern queries over XML data. For example, Amer-Yahia et al. (2004) undertake query relaxation through removal of conditions from XPath expressions; Theobald et al. (2005) support relaxation by expanding queries using vocabulary information drawn from an ontology or thesaurus; Liu et al. (2010) use available XML schemas to relax queries; and Hill et al. (2010) use ontologies such as Wordnet to guide XML query relaxation. Buratti and Montesi (2008) discuss query approximation for XML based on the notion of a cost-based edit distance for transforming one path into another within an XQuery expression, while Almendros-Jimenez et al. (2014) propose a fuzzy approach to XPath query evaluation.

Similar approaches to those developed for XML can be adopted for flexible querying of graph-structured data, and indeed in subsequent sections of this chapter we discuss ontology-based relaxation of graph queries and also edit distance-based ranking of approximate answers to graph queries. However, the techniques proposed for flexibly querying XML generally assume one kind of relationship between entities (parent-child), whereas in graph-structured data there may be numerous relationships, potentially giving rise to higher complexity and diversity in the data and requiring query approximation and relaxation techniques that are able to operate on the relationships referenced within a user's query.

4.1.1 Example: Heterogeneous Data Integration

Much work has been done since the early 1990s in developing architectures and methodologies for integrating biological data (Goble and Stevens 2008). Such integrations are beneficial for scientists by providing them with easy access to more data, leading to more extensive and more reliable analyses and, ultimately, new scientific insights. Traditional data integration methodologies (Batini et al. 1986) require semantic mappings between the different data sources to be initially determined, so that a global integrated schema or ontology can be created through which the data in the sources can then be accessed. This approach means that significant resources for data integration projects must be committed upfront, and an active area of research is how to reduce this upfront effort (Halevy et al. 2006). A general approach adopted is to present initially all of the source data in an unintegrated format, and to provide tools that allow data integrators to incrementally identify semantic relationships between the different data sources and incrementally improve the global schema. Such an approach is termed 'pay-as-you-go' (Sarma and et al. 2008), since the integration effort can be committed incrementally as time and resources allow.

Heterogeneous data integration was identified in Hurtado et al. (2008) as a potential Use Case for flexible query processing over graph data. To illustrate, the In Silico Proteome Integrated Data Environment Resource (ISPIDER) project developed an integrated platform bringing together three independently developed proteomics data sources, providing an integrated global schema and support for distributed queries posed over this (Siepen et al. 2008).[4] The development of the global schema took many months. An alternative approach would have been to adopt a 'pay-as-you-go' integration approach, refining the global ontology by incrementally identifying common concepts between the data sources and integrating these using additional superclasses and superproperties.

For example, the initial ontology may include (amongst others) the following classes arising from three source databases, DB_1, DB_2, DB_3:

- *Peptide$_1$, Protein$_1$, Peptide$_2$, Protein$_2$, Peptide$_3$, Protein$_3$*

For simplicity here, we assume that common concepts are commonly named, and we identify the data source relating to a concept by its subscript. Likewise, it may include (amongst others) the following properties:

- *PepSeq$_i$*, $1 \leq i \leq 3$, each with domain *Peptide$_i$* and range Literal
- *Aligns$_i$*, $1 \leq i \leq 3$, each with domain *Peptide$_i$* and range *Protein$_i$*
- *AccessNo$_i$*, $1 \leq i \leq 3$, each with domain *Protein$_i$* and range Literal

(In proteomics, proteins consist of several peptides, each peptide comprising a sequence of amino acids; hence the properties *PepSeq$_i$* above, in which the amino acid sequence is represented as a Literal. In proteomics experiments, several peptides may result from a protein identification process and each peptide aligns against a set of proteins; hence the properties *Aligns$_i$* above. Each protein is characterised by an Accession Number, c.f. the properties *AccessNo$_i$* above, a textual description, its predicted mass, the organism in which it is found, etc.)

A data integrator may observe some semantic alignments between the above classes and properties and may add the following superclasses and superproperties to the ontology in order to semantically integrate the underlying data extents from the three databases:

- Superclass *Peptide* of classes *Peptide$_i$*, $1 \leq i \leq 3$
- Superclass *Protein* of classes *Protein$_i$*, $1 \leq i \leq 3$
- Superproperty *PepSeq* of properties *PepSeq$_i$*, $1 \leq i \leq 3$, with domain *Peptide* and range Literal
- Superproperty *Aligns* of properties *Aligns$_i$*, $1 \leq i \leq 3$, with domain *Peptide* and range *Protein*
- Superproperty *AccessNo* of properties *AccessNo$_i$*, $1 \leq i \leq 3$, with domain *Protein* and range Literal.

A fragment of this global ontology is shown in Fig. 4.1 (omitting the *AccessNo$_i$* and *AccessNo* properties, and the domain and range information of *PepSeq* and *Aligns*).

---

[4]The example presented here is a simplification of one given in Hurtado et al. (2008).

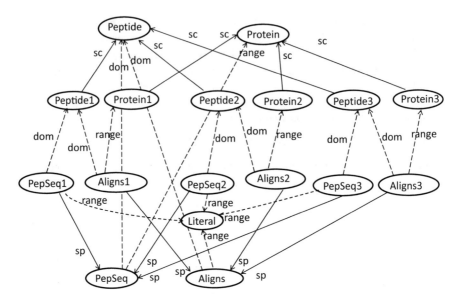

Fig. 4.1 Example ontology

Consider now the following query posed over the global ontology by a user who is only familiar with DB_1:

```
?Y, ?Z <- RELAX(?X,PepSeq1,"ATLITFLCDR"),
          RELAX(?X,Aligns1,?Y),
          RELAX(?Y,AccessNo1,?Z)
```

The syntax used here is that of a conjunctive query comprising one or more *triple patterns* on its right-hand side (RHS)—see Sect. 4.1.2, and zero or more variables on its left-hand side (LHS), which must also appear in the RHS. The entire RHS comprises a *graph pattern*—see Sect. 4.1.2. Variables are distinguished by an initial ?. In its non-relaxed form, this query will return the identifiers and accession numbers of proteins identified in DB_1 through experiments yielding the peptide sequence 'ATLITFLCDR'.

A first level of relaxation of all three triple patterns in the above query results in the following query:

```
?Y, ?Z <- RELAX(?X,PepSeq,"ATLITFLCDR"),
          RELAX(?X,Aligns,?Y),
          RELAX(?Y,AccessNo,?Z)
```

Evaluation of this query will expand the result set to include similar results also from DB_2 and DB_3, without the user needing to have detailed knowledge of their schemas.

In contrast to conventional data integration approaches, this kind of incremental integration coupled with flexible querying requires less upfront integration effort,

Fig. 4.2 RDFS inference rules

$$\text{Subproperty (1)} \frac{(a,sp,b)(b,sp,c)}{(a,sp,c)} \quad (2) \frac{(a,sp,b)(x,a,y)}{(x,b,y)}$$

$$\text{Subclass (3)} \frac{(a,sc,b)(b,sc,c)}{(a,sc,c)} \quad (4) \frac{(a,sc,b)(x,type,a)}{(x,type,b)}$$

$$\text{Typing (5)} \frac{(a,dom,c)(x,a,y)}{(x,type,c)} \quad (6) \frac{(a,range,d)(x,a,y)}{(y,type,d)}$$

allows a more exploratory approach to query answering, and does not require the user to have comprehensive knowledge of the entire global schema.

4.1.2 Theoretical Foundations of Query Relaxation

Hurtado et al. (2008) studied query relaxation in the setting of the RDF/S data model and showed that query relaxation can be naturally formalised using *RDFS entailment*. The entailment was characterised by the derivation rules given in Fig. 4.2, grounded in the semantics developed in Gutierrez et al. (2004) and Hayes (2004), and encompassing a fragment of the overall set of RDFS entailment rules known as ρDF (Munoz et al. 2007).

In the setting of Hurtado et al. (2008), an *ontology* K is a directed graph (N_K, E_K) where each node in N_K represents either a class or a property, so $N_K = classNodes(N_K) \cup propertyNodes(N_K)$; and each edge in E_K is labelled with a symbol from the set $\{sc, sp, dom, range\}$. These edge labels encompass a fragment of the RDFS vocabulary: rdfs:subClassOf, rdfs:subPropertyOf, rdfs:domain, rdfs:range, respectively.

In the accompanying data graph $G = (N, E)$, each node in N represents an instance or a class and each edge in E a property. The intersection of N and N_K is contained in $classNodes(N_K)$. The predicate $type$, representing the RDF vocabulary rdf:type, can be used in E to connect an instance of a class to a node representing that class.

Pairwise disjoint sets U and L of URIs and literals are assumed, respectively. Also assumed is an infinite set V of *variables*, disjoint from U and L. We abbreviate any union of the sets U, L and V by concatenating their names, e.g. $UL = U \cup L$.

Nodes in N are labelled with constants from UL (blank nodes are not considered in this work, and in any case their use is discouraged for Linked Data). Edges in E are labelled either with $type$ or a with symbol drawn from a finite alphabet Σ such that $type \notin \Sigma$ and $\Sigma \cup \{type\} \subset U$.

An *RDF triple* is a tuple $\langle s, p, o \rangle \in U \times U \times (U \cup L)$, where s is the subject, p the predicate and o the object of the triple. A *triple pattern* is a tuple $\langle x, p, y \rangle \in UV \times UV \times UVL$. A *graph pattern* is a set of triple patterns. Given a triple pattern t (graph pattern P), $vars(t)$ ($vars(P)$) is the set of variables occurring in it.

An *RDF/S graph* $I = (N_I, E_I)$ is the union of an ontology graph $K = (N_K, E_K)$ and a data graph $G = (N, E)$, i.e. $N_I = N_K \cup N$ and $E_I = E_K \cup E$.

Fig. 4.3 Additional rules for computing the extended reduction of an ontology

$$(e1)\ \frac{(b,dom,c)(a,sp,b)}{(a,dom,c)}\quad (e2)\ \frac{(b,range,c)(a,sp,b)}{(a,range,c)}$$

$$(e3)\ \frac{(a,dom,b)(b,sc,c)}{(a,dom,c)}\quad (e4)\ \frac{(a,range,b)(b,sc,c)}{(a,range,c)}$$

An RDF/S graph I_1 *entails* an RDF/S graph I_2, denoted $I_1 \models_{RDFS} I_2$, if I_2 can be derived by applying the rules in Fig. 4.2 iteratively to I_1.

The *closure* of an RDF/S graph I under these rules is denoted by $cl(I)$. Given an RDF/S graph I, query evaluation takes place on the graph given by restricting $cl(I)$ to nodes in $N \cup classNodes(N_K)$ and edges with labels in $\Sigma \cup \{type\} \cup propertyNodes(N_K)$. Each such edge is viewed as an RDF triple for the purposes of query evaluation.

In order to apply relaxation to queries, the subgraphs of the ontology K induced by edges labelled sc and sp need to be *acyclic*, so that an unambiguous cost can be assigned to a relaxed query. Moreover, K must be equal to its *extended reduction*, $extRed(K)$, which is computed as follows:

(a) Compute $cl(K)$
(b) Apply the rules of Fig. 4.3 in reverse until no more rules can be applied (applying a rule in reverse means deleting a triple deducible by the rule)
(c) Apply rules 1 and 3 of Fig. 4.2 in reverse until no more rules can be applied

Requiring that $K = extRed(K)$ allows *direct relaxations* to be applied to queries (see below), which correspond to the 'smallest' possible relaxation steps. This in turn allows an unambiguous cost to be associated with relaxed queries, so that query answers can be returned to users incrementally in order of increasing cost (see Hurtado et al. (2008) for a detailed discussion).

Following the terminology of Hurtado et al. (2008), a triple pattern $\langle x, p, y \rangle$ *directly relaxes* to a triple pattern $\langle x', p', y' \rangle$ with respect to an ontology $K = extRed(K)$, denoted $\langle x, p, y \rangle \prec \langle x', p', y' \rangle$, if $vars(\langle x, p, y \rangle) = vars(\langle x', p', y' \rangle)$ and $\langle x', p', y' \rangle$ is derived from $\langle x, p, y \rangle$ by applying some rule i, $1 \leq i \leq 6$, from Fig. 4.2.

A triple pattern $\langle x, p, y \rangle$ *relaxes to* a triple pattern $\langle x', p', y' \rangle$, denoted $\langle x, p, y \rangle \leq \langle x', p', y' \rangle$, if there is a sequence of direct relaxations that derives $\langle x', p', y' \rangle$ from $\langle x, p, y \rangle$. The *relaxation cost* of deriving $\langle x', p', y' \rangle$ from $\langle x, p, y \rangle$ is the minimum cost of applying such a sequence of direct relaxations.

An essential aspect of this approach, which distinguishes it from earlier work on query relaxation, is that the answers to a query are ranked based on how 'closely' they satisfy the query. The notion of ranking is based on a structure called the *relaxation graph*, in which relaxed versions of the original query are ordered from less to more general.

To illustrate, Fig. 4.4 shows the relaxation graphs of two triple patterns:

```
(?X,Aligns1,?Y) and (?X,PepSeq1,"ATLITFLCDR")
```

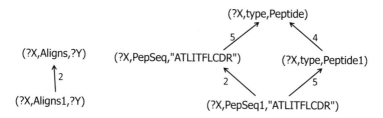

Fig. 4.4 Triple pattern relaxation graphs

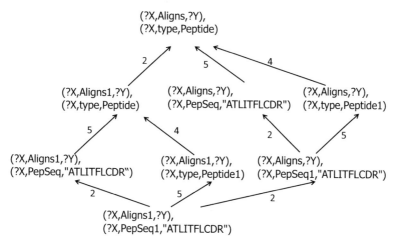

Fig. 4.5 Graph pattern relaxation graph

assuming that K is the ontology of Fig. 4.1. The edges of the relaxation graph are labelled with the rule number from Fig. 4.2 which has been applied to obtain a relaxed triple pattern from one directly below it.

Triple pattern relaxation is generalised to graph pattern relaxation using the notion of the *direct product* of partial orders. The *direct product* of n partial orders $\alpha_1, \alpha_2, \ldots \alpha_n$, denoted $\alpha_1 \otimes \alpha_2 \otimes \ldots \otimes \alpha_n$, is another partial order α such that $(a_1, \ldots a_n) \, \alpha \, (b_1, \ldots b_n)$ if and only if $a_i \, \alpha_i \, b_i$ for all $1 \le i \le n$.

Consider graph patterns consisting of n triple patterns, $t_1, \ldots, t_n$. The graph pattern relaxation relation $\le_n$ is defined as $\le \otimes \le \ldots \otimes \le$ (n times). The direct graph pattern relaxation relation $\prec_n$ is the reflexive and transitive reduction of $\le_n$. The *relaxation graph* of a graph pattern is the directed acyclic graph induced by $\prec_n$.

As an example, consider the graph pattern

```
(?X,Aligns1,?Y),(?X,PepSeq1,"ATLITFLCDR")
```

Figure 4.4 shows the relaxation graphs of its two triple patterns and Fig. 4.5 shows their direct product.

Algorithm `RelaxEval`
Input: a query Q, where $body(Q) = \{t_1, \ldots, t_n\}$; an RDF/S graph I; and an integer *maxLevel*.
Output: the set of ranked answer tuples.

1. $k := 0$, *stillMore* := *true*
2. For each $t_i \in body(Q)$, compute the relaxation graph R_i of t_i up to level *maxLevel*.
3. While ($k \leq maxLevel$ and *stillMore*) do

    a. For each combination $t'_1 \in R_1, \ldots, t'_n \in R_n$ such that $\sum_i level(t'_i, R_i) = k$ do output
       $\pi_H(\texttt{deltaFind}(t'_1, I) \bowtie \ldots \bowtie \texttt{deltaFind}(t'_n, I))$
    b. $k := k + 1$
    c. *stillMore* := exist nodes $t'_1 \in R_1, \ldots, t'_n \in R_n$ such that $\sum_i level(t'_i, R_i) = k$

Fig. 4.6 Algorithm to compute the relaxed answer of a query

Algorithm **RelaxEval** in Fig. 4.6 (from Hurtado et al. 2008) incrementally computes the relaxed answer to a query Q and returns the answers in ranked order, where *maxLevel* is the maximum number of relaxations desired for the evaluation of Q; $body(Q)$ denotes the graph pattern in the RHS of Q; and the set $\texttt{deltaFind}(t'_i, I)$ consists of the triples $\langle s, p, o \rangle \in I$ such that t'_i matches $\langle s, p, o \rangle$ and no triple pattern directly below t'_i in the relaxation graph of t_i matches $\langle s, p, o \rangle$. This algorithm assumes that all direct relaxations of triple patterns have the same cost. We will see later two methods that are able to handle different costs.

Another class of relaxations is also discussed in Hurtado et al. (2008), consisting of relaxations that can be entailed without an ontology, such as dropping triple patterns, replacing constants with variables and breaking join dependencies. We refer the reader to that paper for details of these.

4.2 Beyond Conjunctive Queries: Regular Path Queries

Regular path queries have been proposed by several researchers as a means of assisting users in querying complex or irregular graph data by finding *paths* through the data graph that match a given regular expression over edge labels (Cruz et al. 1987; Mendelzon and Wood 1989, 1995; Fernandez and Suciu 1998).

Consider the same simple data model as introduced above, comprising a directed graph $G = (N, E)$, where each node in N is labelled with a constant and each edge in E is labelled with a symbol drawn from a finite alphabet $\Sigma \cup \{type\}$. Edges can be traversed both from their source to their target node and in reverse, from their target to their source node. The *inverse* of an edge label l, denoted by l^-, is used to specify the reverse traversal of an edge. Let $\Sigma^- = \{l^- \mid l \in \Sigma\}$. If $l \in \Sigma \cup \Sigma^- \cup \{type, type^-\}$, we use l^- to mean the *inverse* of l, that is, if l is a for some $a \in \Sigma \cup \{type\}$, then l^- is a^-, while if l is a^- for some $a \in \Sigma \cup \{type\}$, then l^- is a.

A *regular path query* (RPQ) Q has the form

$$vars \leftarrow (X, R, Y) \tag{4.1}$$

where X and Y are constants or variables, R is a regular expression over $\Sigma \cup \{type\}$, and $vars$ is the subset of $\{X, Y\}$ that are variables. A *regular expression R over* $\Sigma \cup \{type\}$ is defined as follows:

$$R := \epsilon \mid a \mid a^- \mid \_ \mid (R1 \cdot R2) \mid (R1|R2) \mid R^* \mid R^+$$

where ϵ is the empty string, a is any symbol in $\Sigma \cup \{type\}$, '_' denotes the disjunction of all constants in $\Sigma \cup \{type\}$, and the operators have their usual meaning.

A *semipath* (Calvanese et al. 2000) p in $G = (N, E)$ from $v \in N$ to $w \in N$ is a sequence of the form $(v_1, l_1, v_2, l_2, v_3, \ldots, v_n, l_n, v_{n+1})$, where $n \geq 0$, $v_1 = v$, $v_{n+1} = w$ and for each v_i, l_i, v_{i+1} either $v_i \overset{l_i}{\rightarrow} v_{i+1} \in E$ or $v_{i+1} \overset{l_i^-}{\rightarrow} v_i \in E$. A semipath p *conforms* to a regular expression R if $l_1 \cdots l_n \in L(R)$, the language denoted by R.

Given an RPQ Q and graph G, let θ be a matching from $\{X, Y\}$ to nodes of G that maps each constant to itself and such that there is a semipath from $\theta(X)$ to $\theta(Y)$ whose concatenation of edge labels is in $L(R)$. The *answer* of Q on G is the set of tuples $\theta(vars)$ for all such matchings θ.

A *conjunctive regular path query* (CRPQ) Q consisting of n conjuncts has the form

$$Z_1, \ldots, Z_m \leftarrow (X_1, R_1, Y_1), \ldots, (X_n, R_n, Y_n) \tag{4.2}$$

in which each X_i and Y_i, $1 \leq i \leq n$, is a variable or constant, each Z_i, $1 \leq i \leq m$, is a variable appearing in the body of Q, and each R_i, $1 \leq i \leq n$, is a regular expression over $\Sigma \cup \{type\}$.

Given a CRPQ Q and graph G, let θ be a matching from variables and constants of Q to nodes of G such that (i) each constant is mapped to itself, and (ii) there is a semipath from $\theta(X_i)$ to $\theta(Y_i)$ that conforms to R_i, for all $1 \leq i \leq n$. The *answer* of Q on G is the set of m-tuples $\theta(Z_1, \ldots, Z_m)$ for all such matchings θ.

The answer to a CRPQ Q on a graph G can be computed as follows. First find, for each $1 \leq i \leq n$, a binary relation r_i over the scheme (X_i, Y_i) such that tuple $(v, w) \in r_i$ if and only if there is a semipath from node v to node w in G that conforms to R_i, $v = X_i$ if X_i is a constant, and $w = Y_i$ if Y_i is a constant. Then form the natural join of the relations $r_1, \ldots, r_n$ and project over Z_1 to Z_m.

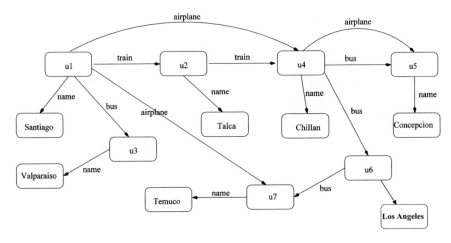

Fig. 4.7 Part of a transport network

4.2.1 Example: Transport Networks

Consider the graph in Fig. 4.7 showing information about a transport network. The nodes of the graph are city identifiers and city names. The edges show direct transport links from one city to another.[5]

Suppose we want to find the cities from which we can travel to city u5 using only airplanes as well as to city u6 using only trains or buses. This can be expressed by the following CRPQ query Q:

```
?X <- (?X, airplane+, u5),
      (?X, (train|bus)+, u6)
```

When Q is evaluated on G, the bindings for ?X generated by the first conjunct are u1, u4, while those for the second conjunct are u1, u2, u4. Hence the answer is u1, u4.

Suppose now that a user who has little knowledge of the structure of the data wishes to find all cities reachable from Santiago by direct flights and poses the following query which makes use of the query approximation operator APPROX that we will discuss in more detail in Sect. 4.2.2:

```
?X <- APPROX (Santiago,airplane,?X)
```

The exact form of this query returns no answers because it does not match the structure of the graph. Inserting name after airplane, to obtain the regular expression airplane.name (at a cost of c_1, say) still returns no answers. Inserting name before airplane, to obtain the regular expression name.airplane.name (at an additional cost of c_1) again returns no answers. Finally, inverting the first name

---

[5]This example is adapted from one in Hurtado et al. (2009b).

label, to obtain `name-.airplane.name` (at an additional cost of c_2, say) returns as answers `Temuco` and `Chillan`, at an overall cost of $2c_1 + c_2$.

Suppose now a user formulates the following query to find cities reachable from Santiago by train, directly or indirectly. The user is also potentially interested in routes combining train and bus, and elects to allow replacement of `train` by `bus` in their query, as well as insertion of `train` and `bus`:

```
?X <- APPROX (Santiago,name-.train+.name,?X)
```

The exact answers to this query are `Talca` and `Chillan`. Replacing one occurrence of `train` by `bus` (at a cost of c_3, say), to obtain the regular expression `name-.bus.train*.name`, returns `Valparaiso` at cost c_3. Inserting `train` after `name-` (at a cost of c_4, say), to obtain `name-.train.bus.train*.name`, returns no more answers. Inserting again `train` after `name-` (at a cost of c_4), to obtain `name-.train.train.bus.train*.name`, returns answers `Concep- cion` and `Los Angeles`, at a cost of $c_3 + 2c_4$. Inserting `bus` before `train*` (at a cost of c_5, say), to obtain `name-.train.train.bus.bus.train*.name` returns answer `Temuco`, at a cost of $c_3 + 2c_4 + c_5$.

4.2.2 Approximate Matching of CRPQs

We have seen above examples of circumstances where *approximate matching* of regular path queries and ranking of query results in terms of how closely they match the original query can help the user find relevant information from unfamiliar, irregular graph data. The work in Hurtado et al. (2009b) discusses how such approximate answers can be computed for CRPQ queries, based on edit operations such as insertions, deletions, inversions, substitutions and transpositions of edge labels being applied to a semipath. A user can specify which of these edit operations should be applied by the system when answering a particular query, and the cost to be assigned when applying each operation, more formally presented as follows.

The *edit distance* from a semipath p to a semipath q is the minimum cost of any sequence of edit operations which transforms the sequence of edge labels of p to the sequence of edge labels of q. The *edit distance* of a semipath p to a regular expression R, $edist(p, R)$, is the minimum edit distance from p to any semipath that conforms to R.

Given a graph G, an RPQ Q of the form (4.1) and a matching θ from variables and constants of Q to nodes in G such that any constant is mapped to itself, the tuple $\theta(vars)$ has *edit distance* $edist(p, R)$ to Q if p is a semipath from $\theta(X)$ to $\theta(Y)$ in G having the minimum edit distance to R of any semipath from $\theta(X)$ to $\theta(Y)$. (Note that if p conforms to R, then $\theta(vars)$ has edit distance 0 to Q.)

The *approximate top-k answer* of Q on G is the list of k tuples $\theta(vars)$ with minimum edit distance to Q, ranked in order of increasing edit distance to Q.

Generalising to CRPQs, given a graph G, a CRPQ Q of the form (4.2), and a matching θ from variables and constants of Q to nodes in G such that any constant is mapped to itself, the tuple $\theta(Z_1, \ldots, Z_m)$ has *edit distance edist*$(p_1, R_1) + \cdots +$ *edist*(p_n, R_n) to Q if each p_i is a semipath from $\theta(X_i)$ to $\theta(Y_i)$ in G having the minimum edit distance to R_i of any semipath from $\theta(X_i)$ to $\theta(Y_i)$. The *approximate top-k answer* of Q on G is the list of k distinct tuples $\theta(Z_1, \ldots, Z_m)$ with minimum edit distance to Q, ranked in order of increasing edit distance to Q.

Since the answers for single conjuncts are ordered by non-decreasing edit distance, pipelined execution of any rank-join operator (see Finger and Polyzotis 2009) can be used to output the answers to a CRPQ Q in order of non-decreasing edit distance.

There are a fixed number of variables in the head of a CRPQ query, so if its conjuncts are acyclic then the evaluation of the approximate top-k answer can be accomplished in polynomial time (see Gottlob et al. 2001; Grahne and Thomo 2001; Hurtado et al. 2009b).

4.3   Combining Approximation and Relaxation in CRPQs

The ideas from the previous two sections can be combined to allow *both* relaxation and approximation of CRPQs, providing their combined flexibility within one query processing framework. This possibility was first explored in Poulovassilis and Wood (2010).

4.3.1   Example: Educational Networks

The L4All system (de Freitas et al. 2008) was developed to support learners in a network of Further and Higher Education institutions in the London region. The system allows users to create and maintain a chronological record of their learning, work and personal episodes—their 'timelines'—with the aim of supporting learners in exploring learning and career opportunities and in planning and reflecting on their learning. Figures 4.8 and 4.9 illustrate a fragment of the data and metadata relating to two users' timelines. The episodes within a timeline have a start and an end date associated with them (for simplicity these are not shown in the figure). Episodes are ordered by their start date—as indicated by edges labelled `next`. There are several types of episode, e.g. `University` and `Work` episodes. Associated with each type of episode are several properties—the figures show just two of these, `qualif[ication]` and `job`.[6]

Suppose that Mary is studying for a BA in English and wishes to find out what possible future career choices there are for her. Timelines may have edges labelled `prereq` between episodes, indicating that the timeline's owner believes

---

[6]This example is adapted from one in Poulovassilis and Wood (2010).

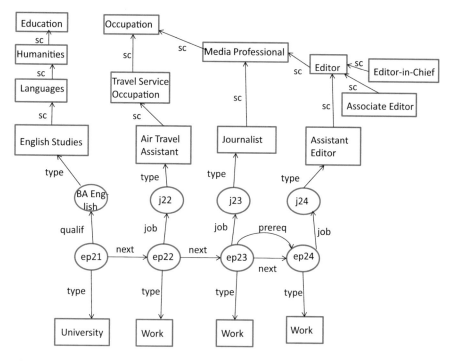

Fig. 4.8 Fragment of data and metadata from Anne's timeline

that undertaking an earlier episode was necessary in order for them to be able to proceed to or achieve a later episode. So Mary might pose this CRPQ query, Q_1:

```
(?E2,?P)<-(?E1,type,University),
          (?E1,qualif.type,EnglishStudies),
          (?E1,prereq+,?E2),
          (?E2,type,Work),
          (?E2,job.type,?P)
```

However, this will return no results even though Anne's timeline in Fig. 4.8 contains information that would be relevant to Mary. This is because, in practice, users may or may not create `prereq` metadata relating to their timelines.

If Mary chooses to allow replacement of the edge label `prereq` in her query by the label `next` (at an edit cost of 1, say), she can submit a variant of Q_1:

```
(?E2,?P)<-(?E1,type,University),
          (?E1,qualif.type,EnglishStudies),
          APPROX (?E1,prereq+,?E2),
          (?E2,type,Work),
          (?E2,job.type,?P)
```

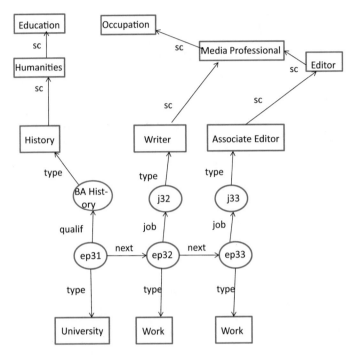

Fig. 4.9 Fragment of data and metadata from Bob's timeline

The regular expression prereq+ can be approximated by the regular expression next.prereq* at edit distance 1 from prereq+, returning the answer

(ep22,AirTravelAssistant)

Mary may judge this not to be relevant and may seek further results, at a further level of approximation. The regular expression next.prereq* can be approximated by next.next.prereq*, at edit distance 2 from prereq+, returning the answers

(ep23,Journalist),(ep24,AssistantEditor)

Mary may judge these as being relevant, and she can then request the system to return the whole of Anne's timeline for her to explore further.

The previous example took as input a starting timeline episode and explored possible future work choices. The next example additionally specifies an end goal and explores how someone might reach this from a given starting point.

Suppose now Mary knows she wants to become an Assistant Editor and would like to find out how she might achieve this, given that she's done an English degree. Mary might pose this query, Q_2:

```
(?E2,?P)<-(?E1,type,University),
           (?E1,qualif.type,EnglishStudies),
           APPROX(?E1,prereq+,?E2),(?E2,job.type,?P)
           APPROX(?E2,prereq+,?Goal),(?Goal,type,Work),
           (?Goal,job.type,AssistantEditor)
```

At edit distance 0 and 1 there are no results from Anne's timeline. At edit distance 2, the answers

```
(ep22,AirTravelAssistant),(ep23,Journalist)
```

are returned, the second of which gives Mary potentially useful information.

Suppose Mary wants to know what other jobs, similar to an Assistant Editor, might be open to her. There are many categories of jobs classified under `Media Professional` but none of these will be matched by her query Q_2 above. She can pose instead query Q_3:

```
(?E2,?P)<-(?E1,type,University),
           (?E1,qualif.type,EnglishStudies),
           APPROX(?E1,prereq+,?E2),(?E2,job.type,?P)
           APPROX(?E2,prereq+,?Goal),(?Goal,type,Work),
           RELAX(?Goal,job.type,AssistantEditor) ,
```

which allows the system to relax `Assistant Editor` to its parent class `Editor`, matching jobs such as `Assistant Editor`, `Associate Editor` etc., as well as in parallel approximating the two instances of `prereq+`. Query results will be returned in increasing overall cost.

As a further extension, suppose another user, Joe, wants to know what jobs similar to being an Assistant Editor might be open to someone who has studied English or a similar subject at university. Joe may pose query Q_4 which is the same as Q_3 above but with RELAX in front of the second conjunct:

```
(?E2,?P)<-(?E1,type,University),
           RELAX(?E1,qualif.type,EnglishStudies),
           APPROX(?E1,prereq+,?E2),(?E2,job.type,?P)
           APPROX(?E2,prereq+,?Goal),(?Goal,type,Work),
           RELAX(?Goal,job.type,AssistantEditor)
```

Suppose Joe sets the cost of relaxing a class to its parent class to 2 and replacing the label `prereq` by the label `next` to 1. Then, the answers produced for query Q_4 from the graphs in Figs. 4.8 and 4.9 are shown in the table below. The first seven columns refer to the answers produced for each of the query conjuncts. For brevity, we do not show the full answer tuples, only the variable instantiations for each conjunct. We also show the relaxation distance (cost), rd, for the second and seventh conjucts, and the edit distance, ed, for the third and fifth conjuncts. In the table, 'Air T.A.' stands for Air Travel Assistant, 'Assist. Ed.' for Assistant Editor and 'Assoc. Ed.' for Associate Editor. The final column shows the overall query answers and their overall distance (d) (which is the sum of the rd and ed values from the second, third, fifth and seventh conjuncts). For greater clarity, the tuples contributing to the first two answers are *italicised* and those contributing to the third answer are shown in **bold**.

?E1	?E1,*rd*	?E1,?E2,*ed*	?E2,?P	?E2,?Goal,*ed*	?Goal	?Goal,*rd*	?E2,?P,*d*
ep21	*ep21,0*	ep23,ep24,0	*ep22,Air T.A.*	*ep23,ep24,0*	ep22	*e24,0*	*ep23,Journalist,2*
ep31	**ep31,4**	*ep21,ep22,1*	*ep23,Journalist*	ep21,ep22,1	*ep23*	**e33,2**	*ep22,Air T.A.,6*
		ep22,ep23,1	ep24,Assist.Ed.	*ep22,ep23,1*	ep24	e23,4	**ep32,Writer,8**
		ep31,ep32,1	**ep32,Writer**	ep31,ep32,1	ep32	e32,4	
		ep32,ep33,1	ep33,Assoc.Ed.	**ep32,ep33,1**	**ep33**	e22,6	
		ep21,ep23,2		ep21,ep23,2			
		ep21,ep24,2		ep21,ep24,2			
		ep31,ep33,2		ep31,ep33,2			

A prototype implementation extending the original L4All system with this flexible querying functionality is described in Poulovassilis et al. (2012). A GUI is provided that allows the user to incrementally build up their query through a forms-based interface, including specifying their preferences for approximation or relaxation to be applied to each subquery. Drop-down menus are used for selecting classes, properties and regular expressions. The CRPQ query is automatically, and incrementally, generated by the system from the user's interactions and preferences. Visualisations are available that allow the user to view at a glance the subqueries they have constructed so far. Query results are displayed one screenful at a time, in increasing distance from the non-approximated, non-relaxed version of the user's query. For each result, an avatar representing the timeline's owner is displayed, as well as their name, the last episode in their timeline matching the user's query, the 'distance' at which this result has been retrieved, and a summary of the timeline's owner and the contents of their timeline. The aim of this summary information is to allow the user to decide if this timeline is relevant for their needs and if they wish to explore it in more detail. These functionalities were evaluated by two Lifelong Learning expert practitioners who gave positive feedback regarding the flexibility of the querying supported and the fact that there is a clear causality between a user's information requirements, as reflected in the query they have constructed, and the results returned by the system.

4.3.2   Automaton-Based Implementation Approach

We now discuss an automaton-based approach to evaluating regular path queries supporting both query approximation and query relaxation. The description is based on that from Poulovassilis and Wood (2010), with some modifications. We refer the reader to Poulovassilis and Wood (2010) and Poulovassilis et al. (2016) for full details.

4.3.2.1 Computing Approximate Answers

Approximate matching of an RPQ query Q with respect to a graph G is achieved by applying edit operations to sequences in $L(R)$. Let q be a sequence in $L(R)$ and l be a label in $\Sigma \cup \Sigma^- \cup \{\texttt{type}, \texttt{type}^-\}$. We assume support for the following edit operations, each at some non-negative cost: *insertion* of l into q, *deletion* of l from q, *substitution* of some label other than l by l in q. The cost of substitution is assumed to be less than the combined cost of insertion and deletion (otherwise the substitution operation would be redundant). The *inversion* operation is achieved through substitution, since this allows some label a in q to be substituted by a^-. The *transposition* operation can be achieved by applying a substitution operation to each of the two labels to be transposed.

Given an RPQ Q with body (X, R, Y) and a graph $G = (N, E)$, the *approximate answer* of Q on G can be computed as follows (the italicised terms are explained in more detail below):

1. A *weighted NFA*, M_R, recognising $L(R)$ is constructed from R.
2. A *query automaton*, M_Q, is constructed from Q.
3. An *approximate automaton*, A_Q, is constructed from M_Q.
4. The *product automaton*, H, of A_Q and G is constructed.
5. One or more shortest path traversals are performed on H in order to find the approximate answer of Q on G.

Definition 4.1 A *weighted non-deterministic finite state automaton* (weighted NFA) M_R recognising $L(R)$ is the same as a normal NFA except that each transition and each final state has a weight associated with it (all of which are initially zero). It can be constructed using Thompson's construction (Aho et al. 1974), which makes use of ϵ-transitions.

Formally, $M_R = (S, \Sigma \cup \Sigma^- \cup \{\texttt{type}, \texttt{type}^-\}, \delta, s_0, S_f, \xi)$, where: S is the set of states; $\Sigma \cup \{type\}$ is the alphabet of edge labels in G; $\delta \subseteq S \times \Sigma \cup \Sigma^- \cup \{\texttt{type}, \texttt{type}^-\} \times \mathbb{N} \times S$ is the transition relation; $s_0 \in S$ is the start state; S_f is the set of final states, initially only consisting of $s_f \in S$; and ξ is the final weight function mapping each state in S_f to a non-negative number (initially, this will be zero for s_f).

The *query automaton* M_Q for Q is M_R with additional annotations on the initial and final states: if X (resp. Y) is a constant c, then s_0 (s_f) is annotated with c; otherwise, s_0 (s_f) is annotated with the symbol $*$ which matches any constant.

Definition 4.2 The *approximate automaton* A_Q for Q is constructed by first constructing an automaton A_R from M_R. Formally, $A_R = (S, \Sigma \cup \Sigma^- \cup \{\texttt{type}, \texttt{type}^-\}, \delta, s_0, S_f, \xi)$, with S, δ, s_0 and S_f initially defined as in Definition 4.1. A_R is then transformed as follows:

- For each transition $(s, a, 0, t) \in \delta$ ($s \neq t$ and $a \in \Sigma \cup \Sigma^- \cup \{\texttt{type}, \texttt{type}^-\}$), the transition (s, ϵ, c_d, t) is added to δ, where c_d is the cost of *deletion*.

- All ϵ-transitions are removed from δ using the method of Droste et al. (2009). The method first computes the ϵ-*closure*, which is the set of pairs of states connected by a sequence of ϵ-transitions along with the minimum summed weight for each such pair. Then for each pair (s, t) with weight w in the ϵ-closure and each transition $(t, b, 0, u) \in \delta$ ($b \neq \epsilon$), a new transition (s, b, w, u) is added to δ. If $t \in S_f$, then s is added to S_f with $\xi[s] = w$ if $\xi[s]$ was previously undefined, or with $\xi[s]$ set to the minimum of $\xi[s]$ and w otherwise.
- For each transition $(s, a, w, t) \in \delta$ and label $b \in \Sigma \cup \Sigma^- \cup \{\texttt{type}, \texttt{type}^-\}$ ($b \neq a$), the transition $(s, b, w + c_s, t)$ is added to δ, where c_s is the cost of *substitution*.
- For each state $s \in S$ and label $a \in \Sigma \cup \Sigma^- \cup \{\texttt{type}, \texttt{type}^-\}$, the transition (s, a, c_i, s) is added to δ, where c_i is the cost of *insertion*.

The *approximate automaton* A_Q for Q is formed from A_R by annotating the initial and final states in A_R with the annotations from the initial and final states, respectively, in M_Q.

Definition 4.3 Let $A_Q = (S, \Sigma \cup \Sigma^- \cup \{\texttt{type}, \texttt{type}^-\}, \delta, s_0, S_f, \xi)$ be an approximate automaton and $G = (N, E)$ be a graph. G can be viewed as an automaton in which each node is both an initial and a final state. The *product automaton* (Mendelzon and Wood 1989), H, of A_Q and G is the weighted automaton $(T, \Sigma \cup \Sigma^- \cup \{\texttt{type}, \texttt{type}^-\}, \sigma, I, F, \xi)$, where $I \subseteq T$ is a set of initial states and $F \subseteq T$ is a set of final states. The set of states T is given by $\{(s, n) \mid s \in S \land n \in N\}$. The set of transitions σ consists of transitions of the form

- $((s, n), a, c, (s', n'))$ if $(s, a, c, s') \in \delta$ and $(n, a, n') \in E$
- $((s, n), a^-, c, (s', n'))$ if $(s, a^-, c, s') \in \delta$ and $(n', a, n) \in E$

The set of initial states I is given by $\{(s_0, n) \mid n \in N\}$. The set of final states F is given by $\{(s_f, n) \mid (s_f, n) \in T \land s_f \in S_f\}$. ξ is the final weight function mapping each state $s \in F$ to a non-negative number. The annotations on initial and final states in H are carried over from the corresponding initial and final states in A_Q.

Having formed the product automaton H, we can now compute the approximate answer of Q on G:

(i) Suppose first that X is a constant v. If $v \notin N$, then the answer is empty. If $v \in N$, we perform a shortest path traversal of H starting from the initial state (s_0, v). Whenever we reach a final state (s_f, n) in H we output n, provided n *matches* the annotation on (s_f, n) (recall that if Y is a constant the annotation on s_f will be that constant, and if Y is a variable the annotation will be the symbol $*$). Node n matches the annotation if the annotation is n or $*$.

(ii) Now suppose X is a variable. In this case, we again perform a shortest path traversal of H, outputting nodes as above, but this time starting from state (s_0, v) for every node $v \in N$.

Two optimisations to this naive traversal to avoid starting at every node of G are described in Selmer et al. (2015). Firstly, if Y is a constant, then (X, R, Y) is transformed to (Y, R^-, X), where R^- is the reversal of R, thus reverting to case (i) above. Otherwise (i.e. both X and Y are variables), we examine the labels on the transitions outgoing from the initial state of A_Q, s_0, we retrieve from G the set of edges (v, l, w) matching these labels, and we perform the shortest path traversal starting from state (s_0, v) for each such node v.

The above evaluation can be accomplished "on-demand" by incrementally constructing the edges of the product automaton H as required, rather than computing the entire graph H, as follows. Three collections are maintained (all initially empty):

- A set visited_R containing tuples of the form (v, n, s), representing the fact that node n of G was visited in state s of A_Q having started the traversal from node v.
- A priority queue queue_R containing tuples of the form (v, n, s, d, f), ordered by non-decreasing values of d, where d is the edit distance associated with visiting node n in state s having started from node v, and f is a flag denoting whether the tuple is 'final' or 'non-final'.
- A list answers_R containing tuples of the form (v, n, d), where d is the smallest edit distance of this answer tuple to Q, ordered by non-decreasing values of d. This list is used to avoid returning an answer (v, n, d') if there is already an answer (v, n, d) with $d \leq d'$.

The evaluation of Q begins by adding to queue_R the initial tuple or tuples $(v, v, s_0, 0, f)$ as detailed in (i) and (ii) above.

Procedure getNext is then called to return the next query answer, in order of non-decreasing edit distance from Q. getNext repeatedly dequeues the first tuple of queue_R, (v, n, s, d, f), adding (v, n, s) to visited_R if the tuple is not a final one, until queue_R is empty. If (v, n, s, d, f) is a final tuple and the answer (v, n, d') has not been generated before for some d', the triple (v, n, d) is returned after being added to answers_R. If (v, n, s, d, f) is not final tuple, we enqueue $(v, m, s', d + d', f))$ for each transition $\xrightarrow{d'} (s', m)$ returned by $\text{Succ}(s, n)$ such that $(v, m, s') \notin \text{visited}_R$. Here, the Succ function returns all transitions $\xrightarrow{d'} (s', m)$ such that there is an edge from (s, n) to (s', m) in H with cost d'. Within Succ, the function $\text{nextStates}(A_Q, s, a)$ returns the set of states in A_Q that can be reached from state s on reading input a, along with the cost of reaching each. If s is a final state, its annotation matches n, and the answer (v, n, d') has not been generated before for some d', then we add the final weight function for s to d, mark the tuple as final, and enqueue the tuple.

Procedure getNext(X, R, Y)

    Input: query conjunct (X, R, Y)
    Output: triple (v, n, d), where v and n are instantiations of X and Y
(1) **while** $nonempty(\text{queue}_R)$ **do**
(2) $(v, n, s, d, f) \leftarrow dequeue(\text{queue}_R)$
(3)   **if** $f \neq \text{'final'}$ **then**
(4)   add (v, n, s) to visited_R
(5)   **foreach** $\xrightarrow{d'} (s', m) \in Succ(s, n)$ s.t. $(v, m, s') \notin \text{visited}_R$ **do**
(6) $enqueue(\text{queue}_R, (v, m, s', d + d', f))$
(7)   **if** s is a final state and its annotation matches n and
               $\nexists d'.(v, n, d') \in \text{answers}_R$ **then**
(8) $enqueue(\text{queue}_R, (v, n, s, d + \xi[s], \text{'final'}))$
(9)   **else**
(10)   **if** $\nexists d'.(v, n, d') \in \text{answers}_R$ **then**
(11)   append (v, n, d) to answers_R
(12)   **return** (v, n, d)
(13) **return** *null*

Function Succ(s, n)

    Input: state s of A_Q and node n of G
    Output: set of transitions which are successors of (s, n) in H
(1) $W \leftarrow \emptyset$
(2) **for** $(n, a, m) \in G$ and $(p, d) \in nextStates(A_Q, s, a)$ **do**
(3)   add the transition $\xrightarrow{d} (p, m)$ to W
(4) **return** W

4.3.2.2 Computing Relaxed Answers

We now describe how the relaxed answer of an RPQ query Q with body (X, R, Y) can be computed, starting from the weighted NFA M_R that recognises $L(R)$. Below we denote by c_i the cost of applying rule i, $i \in \{2, 4, 5, 6\}$, from Fig. 4.2 (since queries and data graphs cannot contain edges labelled sc and sp, rules 1 and 3 are inapplicable to them, although of course they are used in computing the closure of the RDF/S graph).

    Given a weighted automaton $M_R = (S, \Sigma \cup \{\text{type}\}, \delta, s_0, S_f, \xi)$ from which all ϵ-transitions have been removed, and an ontology K such that $K = \text{extRed}(K)$, an automaton $M_R^K = (S', \Sigma \cup \{\text{type}\}, \tau, S_0, S_f', \xi')$ is constructed as described below. The set of states S' includes the states in S as well as any new states defined below. S_0 and S'_f are sets of initial and final states, respectively, with S_0 including the initial state s_0 of M_R, S'_f including the final states S_f of M_R, and both possibly including additional states as defined below. We obtain the *relaxed automaton* of Q with respect to K, M_Q^K, by annotating each state in S_0 and S'_f either with a constant or with $*$ depending on whether X and Y in Q are constants or variables. ξ' is the final weight function mapping states in S'_f to a non-negative number. The transition relation τ includes the transitions in δ as well as any transitions added to τ by the rules defined below. The rules below are repeatedly applied until no further changes to τ and S' can be inferred. The process terminates because of the assumption that the subgraphs of K induced by edges labelled sc and sp are acyclic.

- (rule 2(i)) For each transition $(s, a, d, t) \in \tau$ and triple $(a, sp, b) \in K$, add the transition $(s, b, d + c_2, t)$ to τ.
- (rule 2(ii)) For each transition $(s, a^-, d, t) \in \tau$ and triple $(a, sp, b) \in K$, add the transition $(s, b^-, d + c_2, t)$ to τ.
- (rule 4(i)) For each transition $(s, \texttt{type}, d, t) \in \tau$ such that $t \in S'_f$, t is annotated with c, and $(c, sc, c') \in K$, add to S' a new final state t' annotated with c' (unless there is already such a final state); add a copy of all of t's outgoing transitions to t'; and add the transition $(s, \texttt{type}, d + c_4, t')$ to τ.
- (rule 4(ii)) For each transition transition $(s, \texttt{type}^-, d, t) \in \tau$ such that $s \in S_0$, s is annotated with c, and $(c, sc, c') \in K$, add to S' a new initial state s' annotated with c' (unless there is already such an initial state); add a copy of all of s's incoming transitions to s'; and add the transition $(s', \texttt{type}^-, d + c_4, t)$ to τ.
- (rule 5(i)) For each transition $(s, a, d, t) \in \tau$ such that $t \in S'_f$ and $(a, dom, c) \in K$, add to S' a new final state t' annotated with c (unless there is already such a final state); add a copy of all of t's outgoing transitions to t'; and add the transition $(s, \texttt{type}, d + c_5, t')$ to τ.
- (rule 5(ii)) For each transition $(s, a^-, d, t) \in \tau$ such that $s \in S_0$ and $(a, dom, c) \in K$, add to S' a new initial state s' annotated with c (unless there is already such an initial state); add a copy of all of s's incoming transitions to s'; and add the transition $(s', \texttt{type}^-, d + c_5, t)$ to τ.
- (rule 6(i)) For each transition $(s, a, d, t) \in \tau$ such that $s \in S_0$ and $(a, range, c) \in K$, add to S' a new initial state s' annotated with c (unless there is already such an initial state); add a copy of all of s's incoming transitions to s'; and add the transition $(s', \texttt{type}^-, d + c_6, t)$ to τ.
- (rule 6(ii)) For each transition $(s, a^-, d, t) \in \tau$ such that $t \in S'_f$ and $(a, range, c) \in K$, add to S' a new final state t' annotated with c (unless there is already such a final state); add a copy of all of t's outgoing transitions to t'; and add the transition $(s, \texttt{type}, d + c_6, t')$ to τ.

Having constructed the relaxed automaton M_Q^K, its product automaton with the closure of the graph G is then constructed, and the computation proceeds similarly to cases (i) and (ii) for computing approximate answers above, except that in (i) if X is a class c then the shortest path traversal starts from all initial states (s_0, c') such that c' is a superclass of c. The evaluation can again be accomplished 'on-demand'

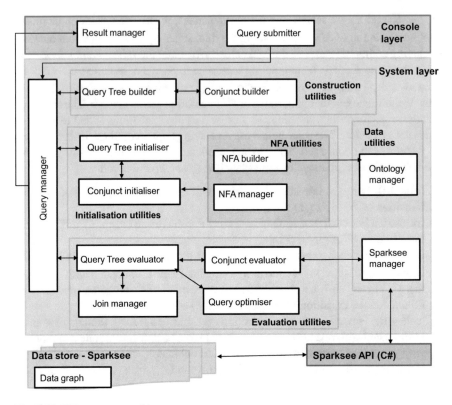

Fig. 4.10 Omega system architecture

by incrementally constructing the edges of the product automaton. The same data structures and algorithms as for computing approximate answers can be used, the only difference being that the Succ function now uses the automaton M_Q^K.

4.3.3 System Architecture and Performance

A prototype implementation of combined approximation and relaxation for CRPQs, called *Omega*, is described in Selmer et al. (2015) and Selmer (2016). Sparksee[7] is used as the data store. The development was undertaken using the Microsoft .NET framework. The system comprises four components (see Fig. 4.10): (1) the *console* layer, to which queries are submitted and which displays the incrementally computed query results; (2) the *system* layer in which query plans are constructed,

---

[7]http://www.sparsity-technologies.com, accessed at 18/6/2015.

optimised and executed; (3) the Sparksee API, which provides an interface for invoking data access methods to the data store; and (4) the data store itself.

Query evaluation commences when *Query submitter* invokes *Query manager*, passing it a CRPQ query that is to be evaluated. *Query manager* invokes *Query Tree builder* to construct the query tree, comprising inner nodes representing join operators and leaf nodes representing individual query conjuncts. *Query Tree builder* calls *Conjunct builder* to construct each leaf node of the query tree. *Query manager* next passes the query tree to *Query Tree initialiser*, which traverses the query tree in a top-down manner, beginning at the root. Whenever *Query Tree initialiser* encounters a leaf node in the query tree, it invokes *Conjunct initialiser* on that conjunct. This in turn invokes *NFA builder* to construct the automaton corresponding to the conjunct's regular expression. If the conjunct is approximated or relaxed, then *NFA manager* is invoked to produce an approximate or relaxed automaton, with the relevant edit or relaxation operators applied. For construction of a relaxed automaton, *NFA manager* interacts with *Ontology manager*, which stores the extended reduction of the ontology. *Query manager* then invokes *Query Tree evaluator*. This first invokes the *Query optimiser* to transform the query tree into its final form for execution (see below for a discussion of optimisation). *Query Tree evaluator* then traverses the optimised query tree, starting from the leftmost leaf node, and proceeding upwards. If the current query tree node is a leaf, the ranked answers for the query conjunct are computed by invoking *Conjunct evaluator*. This module constructs the weighted product automaton, H, of the conjunct's automaton with the (closure of) the data graph G. The construction of H is incremental, with *Conjunct evaluator* invoking *Sparksee manager* to retrieve only those nodes and edges of G that are required in order to compute the next batch of k results (for some predefined value of k, default 100). If the current query tree node is a join, *Query Tree evaluator* works in conjunction with *Join manager* to perform a ranked join of the answers returned thus far by its two children nodes. The join algorithm used is that described in Hurtado et al. (2009b), itself adapted from Ilyas et al. (2004). Once the root of the query tree has been reached, the processing terminates and the list of answers now holds the next k results, ranked by increasing distance. *Query manager* passes this list to *Result manager*, which displays the results in ranked order.

For constructing the automata, use is made of several data structures provided by the C5 Generic Collection library,[8] all of which have an amortised time complexity of O(1) for look-ups and updates:

- HashSet: a set of items (of some type T) implemented as a hash table with linear chaining
- HashedLinkedList: A linked list of items (of some type T) with an additional hash table to optimise item lookups within the list
- HashDictionary: A hash table of typed (key,value) pairs

---

[8]http://www.itu.dk/research/c5, accessed at 18/6/2015.

The `HashDictionary` data structure is used to implement the automata, where the key is an integer representing a 'from' state s, and the value is a `HashedLinkedList` of tuples representing the transitions outgoing from s. The priority queue $queue_R$ is also implemented by a `HashDictionary`. The key is an integer–boolean variable (where the integer portion represents a distance and the boolean portion represents the final or non-final tuples at that distance). The value associated with each key, implemented using a `HashedLinkedList`, comprises tuples of the form (v, n, s, d, f), ordered by increasing values of d, where d is the distance associated with visiting node n in state s having started from node v, and f is a flag denoting whether the tuple is 'final' or 'non-final'. Distinguishing between these two kinds of tuples in the priority queue allows the removal of 'final' tuples to be prioritised, so that answers may be returned earlier.

Readers are referred to Selmer et al. (2015) and Selmer (2016) for further details of the implementation and physical optimisations of the *Omega* system. Those works also report on a performance study of regular path queries with approximation and relaxation on several datasets sourced from the L4All system and from the SIMPLETAX and CORE portions of YAGO (Kasneci et al. 2009). The L4All data graphs used in the performance study were of size up to 220.8 MB for the closure of the data graph while the size of the closure of the YAGO data graph was 1.76 GB. Most of the APPROX and RELAX queries executed quickly on all datasets. However, some of the APPROX queries on YAGO either failed to terminate or did not complete within a reasonable amount of time. This was mainly due to a large number of intermediate results being generated, due to the *Succ* function returning a large number of transitions which are then converted into tuples in *GetNext* and added to $queue_R$. Some optimisations are explored in Selmer et al. (2015) and Selmer (2016) for such queries, enabling several—but not all—of the APPROX queries to execute faster. Future work includes making use of disk-resident data structures for $queue_R$ to guarantee the termination of APPROX queries with large intermediate results, and using knowledge of the graph structure (e.g. to prioritise the evaluation of rarer paths within the graph) to reduce the amount of unnecessary processing. Another promising direction is to identify labels that are rare in the graph and to split the processing of a regular expression into smaller fragments whose first or last label is a rare label, as described in Koschmieder and Leser (2012) (but not for approximated/relaxed queries).

4.4 $SPARQL^{AR}$: Extending SPARQL with Approximation and Relaxation

Relaxation of triple patterns and approximate matching of regular RPQs can be applied to the more pragmatic setting of the SPARQL 1.1 query language (Harris and Seaborne 2013). SPARQL is the predominant language for querying RDF data and, in the latest extension to SPARQL 1.1, it supports RPQs over the RDF graph (known as 'property path queries'). However, it does not support notions of query approximation and relaxation, other than the OPTIONAL operator. Users querying complex RDF datasets may lack full knowledge of the structure of the data, its irregularities, and the URIs used within it. The schemas and URIs used can also evolve over time. This may make it difficult for users to formulate queries that precisely express their information retrieval requirements. Calì et al. (2014) and Frosini et al. (2017) investigate extensions to various fragments of SPARQL 1.1 to allow query approximation and relaxation. These works show that the introduction of the query approximation and query relaxation operators does not increase the complexity class of the language fragments studied, and complexity bounds for several fragments are derived. The extended language is called SPARQL^{AR}.

4.4.1 Example: Flexible Querying of RDF/S Knowledge Bases

Example 4.1 Suppose the user wishes to find the geographic coordinates of the 'Battle of Waterloo' event by posing the following query on the YAGO knowledge base,[9] which is derived from multiple sources such as Wikipedia, WordNet and GeoNames:

```
PREFIX yago:<http://yago-knowledge.org/resource/>
PREFIX rdf:<http://www.w3.org/1999/02/22-rdf-syntax-ns#>
SELECT * WHERE {
  <http://yago-knowledge.org/resource/Battle_of_Waterloo>
  yago:happenedIn/(yago:hasLongitude|yago:hasLatitude)
  ?x }
```

This query uses the *property paths* extension of SPARQL 1.1, including its concatenation (/) and disjunction (|) operators. The above query does not return any answers since YAGO does not store the geographic coordinates of Waterloo.

The user may therefore choose to approximate the triple pattern in their query:

```
SELECT * WHERE {
  APPROX(<http://yago-knowledge.org/resource/Battle_of_Waterloo>
  yago:happenedIn/(yago:hasLongitude|yago:hasLatitude)
  ?x ) }
```

YAGO does store directly the coordinates of the 'Battle of Waterloo' event. So the system can apply an edit operation that deletes happenedIn from the property path, and the resulting query

```
SELECT * WHERE {
  <http://yago-knowledge.org/resource/Battle_of_Waterloo>
  (yago:hasLongitude|yago:hasLatitude)
  ?x }
```

---

[9]http://www.mpi-inf.mpg.de/yago-naga/yago/.

returns the desired answers, showing both high precision and high recall:

```
"4.4"^^<degrees>
"50.68333333333333"^^<degrees>
```

Example 4.2 Consider the following portion $K = (N_K, E_K)$ of the YAGO ontology, where N_K is

$$\{hasFamilyName, hasGivenName, label, actedIn, Actor\}$$

and E_K is

$$\{(hasFamilyName, sp, label), (hasGivenName, sp, label), \\ (actedIn, domain, actor)\}$$

Suppose the user is looking for the family names of all actors who played in the film 'Tea with Mussolini' and poses this query:

```
SELECT * WHERE {
  ?x yago:actedIn <http://yago-knowledge.org/resource/
                   Tea_with_Mussolini> .
  ?x yago:hasFamilyName ?z }
```

The above query returns only four answers, since some actors have only a single name (e.g. Cher), while others have their full name recorded using the `label` property.

The user may choose to relax the second triple pattern in their query in an attempt to retrieve more answers:

```
SELECT * WHERE {
  ?x yago:actedIn <http://yago-knowledge.org/resource/
                   Tea_with_Mussolini> .
  RELAX ( ?x yago:hasFamilyName ?z ) }
```

The system can now replace `hasFamilyName` by `label`, and the resulting query returns the given names of actors in that film recorded through the property `hasGivenName` (hence returning Cher), as well as actors' full names recorded through the property `label`: a total of 255 results.

Example 4.3 Suppose a user wishes to find events that took place in Berkshire in 1643 and poses the following query on YAGO (in the query, we use 'Event' for simplicity but the actual URI is `<wordnet_event_100029378>`):

```
SELECT * WHERE {
  ?x rdf:type Event .
  ?x yago:on "1643-##-##" .
  ?x yago:in "Berkshire" }
```

This query returns no results because there are no property edges named `on` or `in` in YAGO.

The user may choose to approximate the second and third triple patterns of their query:

```
SELECT * WHERE {
    ?x rdf:type Event .
    APPROX ( ?x yago:on "1643-##-##" ) .
    APPROX ( ?x yago:in "Berkshire" ) }
```

The system can now substitute on by happenedOnDate (which does appear in YAGO) and in by happenedIn, giving the following query:

```
SELECT * WHERE {
    ?x rdf:type Event .
    ?x yago:happenedOnDate "1643-##-##" .
    ?x yago:happenedIn "Berkshire" }
```

This still returns no answers, since happenedIn does not connect event instances directly to literals such as "Berkshire".

The user can choose to relax now the third triple pattern of the above query:

```
SELECT * WHERE {
    ?x rdf:type Event .
    ?x yago:happenedOnDate "1643-##-##" .
    RELAX ( ?x yago:happenedIn "Berkshire" )}
```

The system can replace the triple ?x yago:happenedIn "Berkshire" by the triple ?x rdf:type Event, using knowledge encoded in YAGO that the domain of happenedIn is Event, giving the following query, which returns all events recorded as occurring in 1643:

```
SELECT * WHERE {
    ?x rdf:type Event .
    ?x yago:happenedOnDate "1643-##-##" .
    ?x rdf:type Event }
```

Several answers are returned by this query, including the 'Siege of Reading' that happened in 1643 in Berkshire, but also several events that did not happen in Berkshire:

```
<http://yago-knowledge.org/resource/Battle_of_Olney_Bridge>
<http://yago-knowledge.org/resource/Battle_of_Heptonstall>
<http://yago-knowledge.org/resource/Siege_of_Reading>
<http://yago-knowledge.org/resource/Torstenson_War>
<http://yago-knowledge.org/resource/Battle_of_Alton>
<http://yago-knowledge.org/resource/Second_Battle_of_Middlewich>
<http://yago-knowledge.org/resource/Kieft's_War>
```

So the query exhibits better recall than the original query, but possibly low precision.

The user can instead choose to approximate the third triple pattern:

```
SELECT * WHERE {
    ?x rdf:type Event .
    ?x yago:happenedOnDate "1643-##-##" .
    APPROX ( ?x yago:happenedIn "Berkshire" )}
```

The system can now insert the property `label` that connects URIs to their labels, giving the following query:

```
SELECT * WHERE {
    ?x rdf:type Event .
    ?x yago:happenedOnDate "1643-##-##" .
    ?x yago:happenedIn/label "Berkshire" }
```

This query now returns the only event recorded as occurring in 1643 in Berkshire, i.e. the 'Siege of Reading'. It exhibits both better recall than the original query and also high precision.

4.4.2 Query Rewriting-Based Implementation Approach

For specifying the semantics of SPARQL^{AR} queries, we extend the notion of SPARQL query evaluation from returning a set of (exact) mappings to returning a set of mapping/cost pairs of the form $\langle \mu, c \rangle$, where μ is a mapping and c is a non-negative number that indicates the cost of the answers arising from this mapping. Following on from the definitions of sets V, U and L, triples and triple patterns in Sect. 4.1, we have the following definitions (c.f. Pérez et al. 2006):

Definition 4.4 (Mapping) A *mapping* μ from ULV to UL is a partial function $\mu : ULV \rightarrow UL$ such that $\mu(x) = x$ for all $x \in UL$, i.e. μ maps URIs and literals to themselves. The set $var(\mu)$ is the subset of V on which μ is defined. Given a triple pattern $\langle x, z, y \rangle$ and a mapping μ such that $var(\langle x, z, y \rangle) \subseteq var(\mu)$, $\mu(\langle x, z, y \rangle)$ is the triple obtained by replacing the variables in $\langle x, z, y \rangle$ by their image according to μ.

Definition 4.5 (Compatibility and Union of Mappings) Two mappings μ_1 and μ_2 are *compatible* if $\forall x \in var(\mu_1) \cap var(\mu_2)$, $\mu_1(x) = \mu_2(x)$. The *union* of two mappings $\mu = \mu_1 \cup \mu_2$ can be computed only if μ_1 and μ_2 are compatible. The resulting μ is a mapping such that $var(\mu) = var(\mu_1) \cup var(\mu_2)$ and: for each x in $var(\mu_1) \cap var(\mu_2)$, $\mu(x) = \mu_1(x) = \mu_2(x)$; for each x in $var(\mu_1)$ but not in $var(\mu_2)$, $\mu(x) = \mu_1(x)$; and for each x in $var(\mu_2)$ but not in $var(\mu_1)$, $\mu(x) = \mu_2(x)$.

The *union* of two sets of SPARQL^{AR} query evaluation results, $M_1 \cup M_2$, comprises the following set of mapping/cost pairs:

$$\{\langle \mu, c \rangle \mid \langle \mu, c_1 \rangle \in M_1 \text{ or } \langle \mu, c_2 \rangle \in M_2, \text{ with } c = c_1 \text{ if } \nexists c_2.\langle \mu, c_2 \rangle \in M_2, c = c_2 \text{ if }$$
$$\nexists c_1.\langle \mu, c_1 \rangle \in M_1, \text{ and } c = min(c_1, c_2) \text{ otherwise}\}.$$

In Calì et al. (2014) and Frosini et al. (2017) a *query rewriting* approach is adopted for SPARQL^{AR} query evaluation, in which a SPARQL^{AR} query Q is rewritten to a set of SPARQL 1.1 queries for evaluation. We summarise this approach here, refering the reader to those papers for further details.

To keep track of which triple patterns in Q need to be relaxed or approximated, such triple patterns are labelled with A for approximation and R for relaxation. The query rewriting algorithm starts by generating a query Q_0 which returns the exact answer of Q, i.e. ignoring any APPROX and RELAX operators. For each triple pattern $\langle x_i, R_i, y_i \rangle$ in Q_0 labelled with A or R, and each URI p that appears in R_i, a set of new queries is constructed by applying all possible one-step edit operations or relaxation operations to p (these are the 'first-generation' queries). To each such query Q_1 is assigned the cost of applying the edit or relaxation operation that derived it. A new set of queries is constructed by applying a second step of approximation or relaxation to each query Q_1 (the 'second-generation' queries), accumulating summatively the cost of the two edit or relaxation operations applied to obtain each query and assigning this cost to the query. The process continues for a bounded number of generations, accumulating summatively the cost of the sequence of edit or relaxation operations applied to obtain each query in the ith generation. The rewriting process terminates once the cost of all the queries generated in a generation has exceeded a maximum value m.

The overall query evaluation algorithm is defined below, where QRA denotes the Query Rewriting Algorithm and it is assumed that the output set, M, of mapping/cost pairs is maintained in order of increasing cost, e.g. as a priority queue. Ordinary SPARQL query evaluation—denoted $SPARQLeval$ in the algorithm— is applied to each query generated by QRA, in ranked order of the query costs. $SPARQLeval$ takes as input a SPARQL query Q' and a graph G and returns a set of (exact) mappings. The mappings are then assigned the cost of the query Q'. If a mapping is generated more than once, only the one with the lowest cost is retained in M (by the semantics of the union operator, $\cup$, applied to sets of mapping/cost pairs).

Algorithm 7: SPARQL^{AR} flexible query evaluation

---

input : Query Q; maximum cost m; Graph G; Ontology K.
output: List of mapping/cost pairs, M, sorted by cost.
$M := \emptyset$;
foreach $\langle Q', cost \rangle \in QRA(Q, m, K)$ **do**
    foreach $\mu \in SPARQLeval(Q', G)$ **do**
        $M := M \cup \{\langle \mu, cost \rangle\}$
return M;

A formal study of the correctness and termination of the Query Rewriting Algorithm can be found in Frosini et al. (2017) where the Rewriting Algorithm itself is also specified in detail.

4.4.3 System Architecture and Performance

A prototype implementation of SPARQL^{AR} is described in Frosini et al. (2017). The implementation is in Java and Jena is used for the SPARQL query execution.

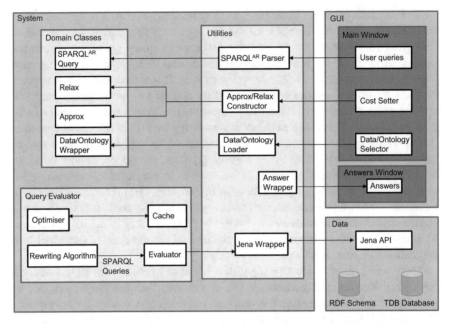

Fig. 4.11 SPARQL^{AR} system architecture

Figure 4.11 illustrates the system architecture, consisting of three layers: the GUI layer, the System layer, and the Data layer.

The GUI layer supports user interaction with the system, allowing queries to be submitted, costs of the edit and relaxation operators to be set, datasets and ontologies to be selected, and query answers to be incrementally displayed to the user.

The System layer is responsible for the processing of the SPARQL^{AR} queries. It comprises three components: the Utilities, containing classes providing the core logic of the system; the Domain Classes, providing classes relating to the construction of SPARQL^{AR} queries; and the Query Evaluator in which query rewriting, optimisation and evaluation are undertaken.

The Data layer connects the system to the selected RDF dataset and ontology using the JENA API. Jena library methods are used to execute SPARQL queries over the RDF dataset and to load the ontology into memory. The RDF datasets are stored as a TDB database[10] and the RDF schema can be stored in multiple RDF formats (e.g. Turtle, N-Triple, RDF/XML).

When a user query is submitted to the GUI, this invokes a method of the *SPARQL^{AR} Parser* to parse the query string and construct an object of the class *SPARQL^{AR} Query*. The GUI also invokes the *Data/Ontology Loader* which creates

---

[10]https://jena.apache.org/documentation/tdb/.

an object of the class *Data/Ontology Wrapper*, and the *Approx/Relax Constructor* which creates objects of the classes *Approx* and *Relax*.

Once these objects have been initialised, they are passed to the Query Evaluator by invoking the *Rewriting Algorithm*. This generates the set of SPARQL queries to be executed over the RDF dataset. The set of queries are passed to the *Evaluator*, which interacts with the *Optimiser* and the *Cache* to improve query performance. Specifically, the answers of parts of queries are computed and stored in the *Cache*, and these answers are retrieved from the *Cache* when the *Evaluator* needs these results. A SPARQL^{AR} query is first split into two parts: the triple patterns which do not have APPROX or RELAX applied to them (the exact part) and those which have (the A/R part). The exact part is first evaluated and the results are cached. The query rewriting algorithm is then applied to the A/R part. Each triple pattern generated is evaluated individually, as also are all possible pairs of triple patterns, and the answers for each evaluation are cached. To avoid memory overflow, an upper limit is placed on the size of the cache. Finally, the overall results of a SPARQL query are obtained by joining subquery results already cached with those obtained by evaluating the rest of the query.

The *Evaluator* uses the *Jena Wrapper* to invoke Jena library methods for executing SPARQL queries over the RDF dataset. The *Jena Wrapper* also gathers the query answers and passes them to the *Answer Wrapper*. Finally, the answers are displayed by the *Answers Window*, in ranked order.

A performance study using data generated from the Lehigh University Benchmark (LUBM)[11] is described in Calì et al. (2014). Three datasets were generated, the largest of which contained 673,416 triples (65 MB). A larger-scale performance study on the YAGO dataset is described in Frosini et al. (2017). YAGO contains over 120 million triples which were downloaded and stored in a Jena TDB database. The size of the TDB database was 9.70 GB, and the nodes of the YAGO graph were stored in a 1.1 GB file.

The overall results show that the evaluation of SPARQL^{AR} queries through a query rewriting approach is promising (see Calì et al. 2014; Frosini et al. 2017 for details). The difference between the execution time of the exact form and the APPROXed/RELAXed forms of the queries is acceptable for queries with fewer than five conjuncts. For most of the other queries that were trialled, the simple caching technique described above also brings down the run times of their APPROXed/RELAXed forms to more reasonable levels. For more complex queries (e.g. involving combinations of Kleene closure and the wildcard symbol, "_", within a property path), more sophisticated optimisation techniques are needed.

Our ongoing work involves investigating optimisations to the query rewriting algorithm, since this can generate a large number of queries. In particular, we are studying the query containment problem for SPARQL^{AR} and how query costs impact on this. For example, for a query $Q = Q_1$ AND Q_2 it is possible to decrease the number of queries generated by the rewriting algorithm if we know

---

[11]http://swat.cse.lehigh.edu/projects/lubm/.

that $Q_1 \subseteq Q_2$, in which case we can evaluate Q_1 rather than Q. Other optimisations under investigation include using statistics about path frequencies in the data graph to reorder the evaluation of triple patterns so as to evaluate first those returning fewer results; and using summaries of the data graph to avoid evaluating subqueries that we know, after evaluation on the graph summary, cannot return any answers. Also planned is a detailed comparison of this query rewriting approach to query approximation and relaxation with the 'native' implementation of Omega described in the previous section.

4.5   Further Topics

4.5.1   User Interaction

In Sect. 4.3.1 we briefly discussed an application-specific prototype that provides a forms-based GUI for incrementally generating CRPQ queries, parts of which can optionally be approximated or relaxed, and for displaying ranked query results to the user. A detailed discussion of that prototype can be found in Poulovassilis et al. (2012). An area of future work identified in that paper was how such systems might provide explanations to the user of how the overall 'distance' of each query result has been derived, based on the application of a sequence of edit and relaxation operations each of some cost specified by the user.

One possible visualisation for such explanations, in a more generic setting, is the Query Graph illustrated in Figs. 4.12 and 4.13, which is based on an 'inverted' version of the relaxation graph for graph patterns discussed in Sect. 4.1. To illustrate, consider Example 4.3 from Sect. 4.1 which is enacted in successive screenshots in the two figures, moving from left-to-right and top-to-bottom. The user begins (Screen 1) by constructing their initial query, which is shown both in the main pane and in the Query Graph panel below. The user then presses the RUN button to run the query. However, no answers are returned (Screen 2). The user elects to edit the second triple pattern, by clicking on that pattern and then on the 'Conj[unct]' button, selecting 'substitution' from a drop-down list of edit operations displayed by the system, and then happenedOnDate from a list of properties suggested by the system (e.g. properties that are known to have domain Event). Screen 3 shows the new query and its distance from the original one, the updated Query Graph and—in the top-left—the edit operations applied so far. The user presses RUN but again no answers are returned. The user elects to edit now the third triple pattern, again selecting 'substitution' from a drop-down list of edit operations, and now happenedIn from the list of properties suggested by the system. Screen 4 shows the new query and the updated Query Graph. The user presses RUN but again no answers are returned. At this point, the user seeks help by clicking on the "?" button and system suggests three alternatives: (i) relaxation of the second triple pattern to ?x rdf:type Event, (ii) relaxation of the third triple pattern

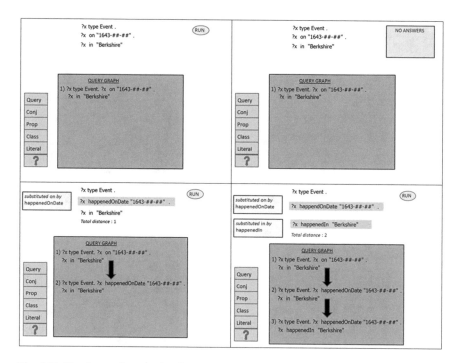

Fig. 4.12 User interaction and visualisation

to ?x rdf:type Event, (iii) insertion of a literal-valued property to follow happenedIn. Suppose the user chooses option (ii). Screen 5 shows the new query and the updated Query Graph (scrolling down now in the expanded Query Graph pane). The user presses RUN and Screen 6 illustrates the results returned (all events taking place in 1643, at any location). The user decides these results are too diverse to be useful and backtracks to Query 3, where the system provides again alternatives (i)–(iii). Suppose the user now chooses option (iii). The system offers a list of literal-valued properties (those with domain Place, or a superclass, on the basis of knowledge that this is the range of happenedIn), and the user selects the property label. Screen 7 shows the new query and the updated Query Graph. The user presses RUN and Screen 8 shows the result returned, which is the one event recorded as occurring in Berkshire in 1643.

Allowing the user to visualise how queries are incrementally generated, what distance is associated with each query, and what results are returned, if any, can help the user decide whether the answers being returned are useful and to try out different edit/relaxation operations. Detailed design, implementation and evaluation of such interactive flexible querying facilities and visualisations for end-users are an area requiring further work.

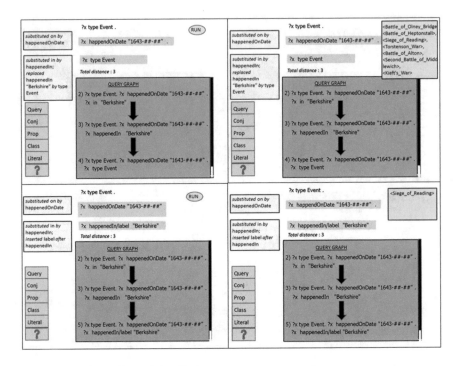

Fig. 4.13 User interaction and visualisation

4.5.2 More Query Flexibility

There are several directions in which the approaches discussed in the previous sections can be extended. One area of ongoing work is to merge the APPROX and RELAX operators into one integrated FLEX operator that *simultaneously* applies edit and relaxation operations to a regular path query. This would allow greater ease of querying for users, in that they would not need to be aware of the ontology structure and to identify which conjuncts of their query may be amenable to relaxation and which to approximation. Another ongoing direction is to extend our languages with lexical and semantic *similarity measures*, in order to allow approximate matching of literals and resources.

To illustrate, suppose a History of Science researcher wishes to find scientists born in London. She is also interested in scientists living in or near London. However, she doesn't know how YAGO records that a person is a scientist, and poses the following query that makes use of the `hasGloss` property linking resources to textual descriptions, seeking to find the word 'scientist' or similar within such descriptions:

```
SELECT * WHERE {
    FLEX (?p yago:wasBornIn London) .
```

```
?p rdf:type ?c . ?c yago:hasGloss ?descr .
FILTER sim (?descr, "scientist") > 0.7}
```

The system can find matches for 'scientist' within values of the `hasGloss` property (e.g. as in the descriptions 'person with advanced knowledge of one or more sciences', 'an expert in the science of economics'), allowing relevant answers to be returned. Use of the FLEX operator also allows the system to substitute `wasBornIn` by `livesIn`, giving additional answers of relevance.

Similarity measures could also be applied by the system to distinguish between different alternatives when an edge label in a regular path query is being substituted by a different label. Having knowledge of the semantic similarly of properties— for example, exploiting dictionaries such as Wordnet—would allow the system to assign a finer-grained cost to edge label substitutions, thereby allowing finer ranking of the top-k answers and increasing their precision.

Similarly, it would be possible to assign a finer ranking to the application of relaxation operations where a specific property is replaced by superproperty, or a specific class by a superclass.

4.5.3 More Query Expressivity

Another direction of work is to extend the expressivity of the query languages beyond conjunctive regular path queries. In this direction, Hurtado et al. (2009a) investigated approximate matching of *extended regular path* (ERP) queries in which the regular expression R in a query conjunct (X, R, Y) can be associated with a path variable P—using the syntax $(X, R : P, Y)$—and these path variables can appear also in the query head, thereby allowing graph paths to be returned to the user as part of the query answer. It was shown that top-k approximate answers can be returned in polynomial time in the size of the graph and the query. Thus, for example, revisiting the transport network example in Sect. 4.2.1, the following query finds cities reachable from Santiago by train, directly or indirectly, or by combinations of other modes of transport, returning also the routes:

```
?P, ?X <- APPROX (u1, train+.name:?P ,?X)
```

The answers returned are:
```
([train,u2,name],Talca),
([train,u2,train,u4,name],Chillan)
```
at distance 0;
```
([bus,u3,name],Valparaiso),
([airplane,u7,name],Temuco),
([airplane,u4,name],Chillan)
```
at distance c_3;
```
([airplane,u4,airplane,u5,name],Concepcion),
([airplane,u4,bus,u5,name],Concepcion),
```

```
([airplane,u4,bus,u6,name],Los Angeles)
```
at distance $c_3 + c_4$; and so forth.

Another application of this kind of flexible querying, where it is useful to return paths in the query results, is described in Poulovassilis et al. (2015), which discusses the analysis of user–system interaction data as arising from exploratory learning environments. The interaction data is stored in Neo4j. Interaction events, and their types, are represented by nodes. Events are linked to their event-type by edges labelled OCCURRENCE_OF while successive events are linked to each other by edges labeled NEXT. Poulovassilis et al. (2015) give examples of how approximate matching of ERP queries over such data can allow pedagogical experts to investigate how students are undertaking exploratory learning tasks, and how feedback messages generated by the system are affecting students' behaviours, with the aim of designing improved support for students. For example, the following query (expressed in Neo4j's Cypher language[12]) finds pairs of events x, y such that x is an intervention (i.e. a message) generated by the system and y is the user's next action; the path between x and is returned, through the variable p, as are the event-type of x and y, through the variables v and w.

```
MATCH (x:Event)-[:OCCURRENCE_OF]->
               (v:EventType {event_cat:"intervention"}),
        p = (x:Event)-[:NEXT]->(y:Event),
        (y:Event)-[:OCCURRENCE_OF]->(w:EventType)
RETURN v.event_type as start_node_type,
        extract(n IN nodes(p) | n.id_fltask) as path_node_ids,
        w.event_type as end_node_type
```

Results returned include:

start_node_type	path_node_ids	end_node_type
highMessage	["344509","344510"]	ClickButton
highMessage	["344519","344520"]	ClickButton
highMessage	["344522","344523"]	ClickButton
highMessage	["344714","344715"]	ClickButton
highMessage	["344760","344761"]	ClickButton

If query approximation were supported in Cypher, then applying APPROX to the subquery `(x:Event)-[:NEXT]->(y:Event)` above, and enabling just insertion of edge labels, would generate the subquery `(x:Event)-[:NEXT*2]->`

---

[12]http://neo4j.com, accessed at 18/6/2015.

(y:Event). Evaluation of the new query would return answers such as the following, all at distance 1 from the original query:

start_node_type	path_node_ids	end_node_type
highMessage	["344509","344510","346027"]	ClickButton
highMessage	["344519","344520","344521"]	PlatformEvent
highMessage	["344522","344523","344712"]	ClickButton
highMessage	["344714","344715","344716"]	ClickButton
highMessage	["344760","344761","344762"]	PlatformEvent

Following this, the subquery (x:Event)-[:NEXT*2]->(y:Event) could be automatically approximated again to (x:Event)-[:NEXT*3]->(y:Event). Evaluation of the new query would return answers such as the following, now at edit distance 2 from the original query:

start_node_type	path_node_ids	end_node_type
highMessage	["344509","344510","346027","346028"]	FractionGenerated
highMessage	["344519","344520","344521","344522"]	highMessage
highMessage	["344522","344523","344712","344713"]	FractionGenerated
highMessage	["344714","344715","344716","344717"]	FractionChange
highMessage	["344760","344761","344762","344763"]	highMessage

By this point, the pedagogical expert is able to see that some high-level interruption messages ('highMessage') are leading students towards productive behaviours, such as generating a fraction—lines 1 and 3, or changing a fraction—line 4 (the specific learning environment to which this data relates aims to teach young learners about fractions). However, other messages are just resulting in more messages being generated by the system (lines 2 and 5), which may lead experts to explore the data further (e.g. to retrieve the messages associated with events 344519 and 344760) and possibly reconsider this part of the system's design.

4.6   Related Work

A general overview of graph databases from the perspectives of graph character-istics, graph data management, applications and benchmarking can be found in Larriba-Pey et al. (2014). The work described in this chapter has considered only the simple graph data model introduced in Sect. 4.1, and a broader survey of graph data models can be found in Angles and Gutierrez (2008). Likewise, a survey of graph query languages can be found in Wood (2012), and we focus here on languages that support RPQs and on flexible query processing for graph-structured data.

Using regular expressions to specify path queries on graph-structured data has been studied for nearly 30 years, being introduced in the languages G, G+ and

Graphlog (Cruz et al. 1987; Consens and Mendelzon 1989; Mendelzon and Wood 1989, 1995) and taken up in several languages for other semi-structured data models (Abiteboul et al. 1997; Fernandez et al. 2000; Buneman et al. 2000). More recently, CRPQs are supported in NAGA (Kasneci et al. 2009), SPARQLeR (Kochut and Janik 2007), PSPARQL (Alkhateeb et al. 2009), G-SPARQL (Sakr et al. 2012) and SPARQL 1.1 (Harris and Seaborne 2013). Cypher, the declarative query language supported by the Neo4j graph DBMS, also supports a restricted form of regular path queries. The nSPARQL language (Pérez et al. 2008) extends SPARQL with *nested* regular expressions and shows that these enable query answers that encompass the semantics of the RDFS vocabulary by direct graph traversal, without materialising the closure of the graph. In addition to the automaton-based approach described in Sect. 4.3, other approaches proposed for evaluating (exact) CRPQs include translation into Datalog or recursive SQL (Consens and Mendelzon 1993; Wood 2012; Dey et al. 2013), search-based processing (Fan et al. 2011; Koschmieder and Leser 2012) and reachability indexing (Gubichev et al. 2013).

Recent work in the WAVEGUIDES project is investigating cost-based optimisation for SPARQL 1.1, focusing in the first instance on query optimisation for property paths (Yakovets et al. 2015), and this has potential application in the optimisation of approximated/relaxed CRPQs as well.

Early work on flexible querying for semi-structured data was undertaken by Kanza and Sagiv (2001), who considered matchings returning paths whose set of edge labels contain those appearing in the query; such semantics can be captured by transposition and insertion edit operations on edge labels. More generally, Grahne and Thomo (2001, 2006) explored approximate matching of single-conjunct regular path queries, using a weighted regular transducer to perform transformations to RPQs for approximately matching semi-structured data. This approach was extended in Hurtado et al. (2009b) to CRPQs. In other work, Grahne et al. (2007) introduced *preferential* RPQs where users can specify the relative importance of symbols appearing in the query by annotating them with weights.

The work in Barcelo et al. (2010, 2012) extends CRPQs to allow comparisons between path variables within the bodies of queries, as well as allowing path variables to appear in query heads, calling this extension *extended conjunctive regular path queries* (ECRPQs) (but not considering flexible querying). The work in Libkin and Vrgoc (2012) extends CRPQs to include manipulation also of the data values associated with nodes along a path. Other extensions to CRPQs are discussed in Wood (2012), for example with aggregation functions such as *count*, *sum*, *max*, *min* to allow finding properties of graphs that are useful for network analysis (e.g. in/out-degree of nodes, length of shortest paths between nodes, graph diameter). Extending these more expressive graph query languages with flexible querying capabilities is an open area. Also open is extending graph query languages for more complex graph models (e.g. property graphs, hyperedges, hypernodes— see Angles and Gutierrez 2008) with flexible queries.

There have been several proposals for flexibly querying Semantic Web data using *similarity measures* to retrieve additional relevant answers. For example, in iSPARQL (Kiefer et al. 2007) similarity measures are applied to resources; in

Hogan et al. (2012) similarity functions are applied to constants such as strings and numeric values; and in De Virgilio et al. (2013) a structural similarity approach is proposed that exploits the graph structure of the data. In other work, ontology-driven similarity measures are developed, using an RDFS ontology to retrieve additional answers and assign a score to them (Huang et al. 2008; Huang and Liu 2010; Reddy and Kumar 2010).

In Mandreoli et al. (2009) knowledge of the semantic relationships between graph nodes is used for approximate query matching, and Cedeno and Candan (2011) describe a framework for cost-aware querying of weighted RDF data through predicates that express flexible paths between nodes. Elbassuoni et al. (2009, 2011) propose extending SPARQL with keyword search capabilities, together with IR-style ranking of query answers. In Yang et al. (2014), a set of transformation functions are used to map attributes of nodes and edges appearing in a graph query to matches in the data graph, and a ranking model for query answers is learnt using automatically generated training instances and the query log.

Dolog et al. (2006, 2009) consider relaxing queries on RDF data based on user preferences; user preferences mined from the query log are also used for query relaxation in Meng et al. (2008); and flexible querying using preferences expressed as fuzzy sets is investigated in Buche et al. (2009).

Approximate *graph matching* has also been much studied (Zhang et al. 2010; Zhu et al. 2011; Zou et al. 2011; Fan et al. 2013; Ma et al. 2014), including adding regular expressions as edge constraints on the graph patterns to be matched (Fan et al. 2011) and ontology-based subgraph querying (Wu et al. 2013). This work has synergies with the flexible querying processing approaches discussed in this chapter, since the algorithms proposed could potentially be leveraged for improved query performance of approximated/relaxed CRPQs: this is currently an open area of research.

4.7 Concluding Remarks

We have given an overview of motivations, applications and implementation techniques for extending graph query languages with relaxation and approximation. Along the way we have highlighted directions of ongoing work, relating to providing additional flexibility through similarity matching, designing further logical and physical optimisations, and conducting more extensive performance studies. On the theory front, future work involves investigating the query containment problem for $SPARQL^{AR}$ and the complexity implications of extending more expressive query languages with relaxation and approximation features. On the usability front, further work is required on designing user interfaces that allow users to control and visualise how flexible queries are incrementally generated and evaluated, so as to be able to decide whether the answers being returned are useful and to try out alternative relaxations or approximations.

Acknowledgements The author gratefully thanks Andrea Calì, Riccardo Frosini, Sergio Gutierrez-Santos, Carlos Hurtado, Manolis Mavrikis, Petra Selmer, Alex Wollenschlaeger and Peter Wood for our collaboration in the work described here.

References

Abiteboul S, Quass D, McHugh J, Widom J, Wiener J (1997) The LOREL query language for semistructured data. Int J Digit Libr 1(1):68–88

Aho AV, Hopcroft JE, Ullman JD (1974) The design and analysis of computer algorithms. Addison-Wesley, Reading

Alkhateeb F, Baget J, Euzenat J (2009) Extending SPARQL with regular expression patterns (for querying RDF). J Web Semantics 7(2):57–73

Almendros-Jimenez J, Luna A, Moreno G (2014) Fuzzy XPath queries in XQuery. In: Proceedings of OTM 2014, pp 457–472

Amer-Yahia S, Lakshmanan LVS, Pandit S (2004) FleXPath: flexible structure and full-text querying for XML. In: Proceedings of ACM SIGMOD 2004, pp 83–94

Angles R, Gutierrez C (2008) Survey of graph database models. ACM Comput Surv 40(1):1–39

Ayers R (1997) Databases for criminal intelligence analysis: knowledge representation issues. AI Soc 11:18–35

Babcock B, Chaudhuri S, Das G (2003) Dynamic sample selection for approximate query processing. In: Proceedings of ACM SIGMOD 2003, pp 539–550

Barcelo P, Hurtado CA, Libkin L, Wood PT (2010) Expressive languages for path queries over graph-structured data. In: Proceedings of PODS 2010, pp 3–14

Barcelo P, Libkin L, Lin AW, Wood PT (2012) Expressive languages for path queries over graph-structured data. ACM Trans Database Syst 37(4):1–46

Batini C, Lenzerini M, Navathe SB (1986) A comparative analysis of methodologies for database schema integration. ACM Comput Surv 18(4):323–364

Bordogna G, Psaila G (2008) Customizable flexible querying in classical relational databases. In: Handbook of research on fuzzy information processing in databases. IGI Global, Hershey, pp 191–217

Bosc P, Pivert O (1992) Some approaches for relational databases flexible querying. J Intell Inf Syst 1(3):323–354

Bosc P, Hadjali A, Pivert O (2009) Incremental controlled relaxation of failing flexible queries. J Intell Inf Syst 33(3):261–283

Bray T et al (eds) (2008) Extensible markup language (XML) 1.0, W3C Recommendation

Buche P, Dibie-Barthelemy J, Chebil H (2009) SPARQL querying of web data tables driven by an ontology. In: Proceedings of FQAS 2009

Buneman P, Fernandez M, Suciu D (2000) A query language and algebra for semistructured data based on structural recursion. VLDB J 9(1):76–110

Buratti G, Montesi D (2008) Ranking for approximated XQuery full-text queries. In: Proceedings of BNCOD 2008, pp 165–176

Calì A, Frosini R, Poulovassilis A, Wood PT (2014) Flexible querying for SPARQL. In: Proceedings of ODBASE 2014 (OTM Conferences), pp 473–490

Calvanese D, Giacomo GD, Lenzerini M, Vardi MY (2000) Containment of conjunctive regular path queries with inverse. In: Proceedings of KR 2000, pp 176–185

Cedeno J, Candan KS (2011) R2DF framework for ranked path queries over weighted RDF graphs. In: Proceedings of WIMS 2011

Chakrabarti K, Garofalakis M, Rastogi R, Shim K (2001) Approximate query processing using wavelets. VLDB J 10(2–3):199–223

Chen AC, Gao S, Karampelas P, Alhajj R, Rokne J (2011) Finding hidden links in terrorist networks by checking indirect links of different sub-networks. In: Counterterrorism and open source intelligence, pp 143–158

Chu W, Yang H, Chiang K, Minock M, Chow G, Larson C (1996) CoBase: a scalable and extensible cooperative information system. J Intell Inf Syst 6(2/3):223–259

Consens M, Mendelzon AO (1989) Expressing structural hypertext queries in GraphLog. In: Proceedings of ACM hypertext 1989, pp 269–292

Consens M, Mendelzon AO (1993) Low complexity aggregation in Graphlog and Datalog. Theor Comput Sci 116(1–2):95–116

Cruz IF, Mendelzon AO, Wood PT (1987) A graphical query language supporting recursion. In: Proceedings of SIGMOD 1987, pp 323–330

de Freitas S, Harrison I, Magoulas G, Papamarkos G, Poulovassilis A, van Labeke N, Mee A, Oliver M (2008) L4All: a web-service based system for lifelong learners. In: The learning grid handbook: concepts, technologies and applications. The future of learning, vol 2. IOS Press, Amsterdam

De Virgilio R, Maccioni A, Torlone R (2013) A similarity measure for approximate querying over RDF data. In: Proceedings of EDBT/ICDT 2013 workshops, pp 205–213

Deo N (2004) Graph theory with applications to engineering and computer science. PHI Learning, New Delhi

Dey S, Cuevas-Vicenttin V, Kohler S, Gribkoff E (2013) On implementing provenance-aware regular path queries with relational query engines. In: Proceedings of EDBT 2013, pp 214–223

Dolog P, Stuckenschmidt H, Wache H (2006) Robust query processing for personalized information access on the semantic web. In: Proceedings of FQAS 2006

Dolog P, Stuckenschmidt H, Wache H, Diederich J (2009) Relaxing RDF queries based on user and domain preferences. J Intell Inf Syst 33(3):239–260

Droste M, Kuich W, Vogler H (2009) Handbook of weighted automata. Springer, Berlin

Eckhardt A, Hornicak E, Vojtas P (2011) Evaluating top-k algorithms with various sources of data and user preferences. In: Proceedings of FQAS 2011, pp 258–269

Elbassuoni S, Ramanath M, Schenkel R, Sydow M, Weikum G (2009) Language model-based ranking for queries on RDF-graphs. In: Proceedings of CIKM 2009, pp 977–986

Elbassuoni S, Ramanath M, Weikum G (2011) Query relaxation for entity-relationship search. In: Proceedings of ESWC 2011 (Part 2)

Fan W, Li J, Ma S, Tang N, Wu Y (2011) Adding regular expressions to graph reachability and pattern queries. In: Proceedings of ICDE 2011, pp 39–50

Fan W, Wang X, Wu Y (2013) Diversified top-k graph pattern matching. PVLDB 6(13):1510–1521

Fernandez M, Suciu D (1998) Optimizing regular path expressions using graph schemas. In: Proceedings of ICDE 1998, pp 14–23

Fernandez M, Florescu D, Levy A, Suciu D (2000) Declarative specification of web sites with strudel. VLDB J 9(1):38–55

Finger J, Polyzotis N (2009) Robust and efficient algorithms for rank join evaluation. In: Proceedings of SIGMOD 2009

Fink R, Olteanu D (2011) On the optimal approximation of queries using tractable propositional languages. In: Proceedings of ICDT 2011, pp 174–185

Frosini R, Calì A, Poulovassilis A, Wood PT (2017) Flexible query processing for SPARQL. Semantic Web 8(4):533–563

Galindo J, Medina J, Pons O, Cubero C (1998) A server for fuzzy SQL queries. In: Proceedings of FQAS 1998, pp 164–174

Goble CA, Stevens R (2008) State of the nation in data integration for bioinformatics. J Biomed Inform 41(5):687–693

Gottlob G, Leone N, Scarcello F (2001) The complexity of acyclic conjunctive queries. J ACM 43(3):431–498

Grahne G, Thomo A (2001) Approximate reasoning in semi-structured databases. In: Proceedings of KRDB 2001

Grahne G, Thomo A (2006) Regular path queries under approximate semantics. Ann Math Artif Intell 46(1–2):165–190

Grahne G, Thomo A, Wadge WW (2007) Preferentially annotated regular path queries. In: Proceedings of ICDT 2007, pp 314–328

Gubichev A, Bedathur S, Seufert S (2013) Sparqling kleene: fast property paths in rdf-3x. In: Proceedings of 1st international workshop on graph data management experiences and systems (GRADES'13)

Gutierrez C, Hurtado C, Mendelzon AO (2004) Foundations of Semantic Web Databases. In: Proceedings of PODS 2004, pp 95–106

Halevy A, Rajaraman A, Ordille J (2006) Data integration: the teenage years. In: Proceedings of VLDB 2006, pp 9–16

Harris S, Seaborne A (eds) (2013) SPARQL 1.1 Query Language, W3C Recommendation

Hayes P (ed) (2004) RDF Semantics, W3C Recommendation

Heer J, Agrawala M, Willett M (2008) Generalized selection via interactive query relaxation. In: Proceedings of CHI 2008, pp 959–968

Hill J, Torson J, Guo B, Chen Z (2010) Toward ontology-guided knowledge-driven XML query relaxation. In: Proceedings of 2nd international conference on computational intelligence, modelling and simulation (CIMSiM) 2010, pp 448–453

Hogan A, Mellotte M, Powell G, Stampouli D (2012) Towards fuzzy query relaxation for RDF. In: Proceedings of ISWC 2012, pp 687–702

Huang H, Liu C (2010) Query relaxation for star queries on RDF. In: Proceedings of WISE 2010, pp 376–389

Huang H, Liu C, Zhou X (2008) Computing relaxed answers on RDF databases. In: Proceedings of WISE 2008, pp 163–175

Hurtado CA, Poulovassilis A, Wood PT (2008) Query relaxation in RDF. J Data Semantics X:31–61

Hurtado CA, Poulovassilis A, Wood PT (2009a) Finding top-k approximate answers to path queries. In: Proceedings of FQAS 2009, pp 465–476

Hurtado CA, Poulovassilis A, Wood PT (2009b) Ranking approximate answers to semantic web queries. In: Proceedings of ESWC 2009, pp 263–277

Ilyas I, Aref W, Elmagarmid A (2004) Supporting top-k join queries in relational databases. VLDB J 13:207–221

Ioannidis Y, Poosala V (1999) Histogram-based approximation of set-valued query-answers. In: Proceedings of VLDB 1999, pp 174–185

Kanza Y, Sagiv Y (2001) Flexible queries over semistructured data. In: Proceedings of ACM PODS 2001, pp 40–51

Kasneci G, Ramanath M, Suchanek F, Weikum G (2009) The YAGO-NAGA approach to knowledge discovery. ACM SIGMOD Rec 37(4):41–47

Kiefer C, Bernstein A, Stocker M (2007) The fundamentals of iSPARQL: a virtualtriple approach for similarity-based semantic web tasks. In: Proceedings of ISWC 2007

Kochut K, Janik M (2007) Extended SPARQL for semantic association discovery. In: Proceedings of ESWC 2007, pp 145–159

Koschmieder A, Leser U (2012) Regular path queries on large graphs. In: Proceedings of SSDBM 2012, pp 177–194

Lacroix Z, Murthy H, Naumann F, Raschid L (2004) Links and paths through life sciences data sources. In: Proceedings of DILS 2004, pp 203–211

Larriba-Pey J, Martinez-Bazan N, Dominguez-Sal D (2014) Introduction to graph databases. In: Proceedings of reasoning web 2014, pp 171–194

Leser U, Trissl S (2009) Graph management in the life sciences. In: Encyclopedia of database systems, pp 1266–1271

Libkin L, Vrgoc D (2012) Regular path queries on graphs with data. In: Proceedings of ICDT 2012, pp 74–85

Liu C, Li J, Yu J, Zhou R (2010) Adaptive relaxation forquerying heterogeneous XML data sources. Inf Syst 35(6):688–707

Ma S, Cao Y, Fan W, Huai J, Wo T (2014) Strong simulation: capturing topology in graph pattern matching. ACM Trans Database Syst 39(1):1–46

Mandreoli F, Martoglia R, Villani G, Penzo W (2009) Flexible query answering on graph-modeled data. In: Proceedings of EDBT 2009, pp 216–227

Martin MS, Gutierrez C, Wood PT (2011) A social networks query and transformation language. In: Proceedings of AMW 2011, pp 631–646

Mendelzon AO, Wood PT (1989) Finding regular simple paths in graph databases. In: Proceedings of VLDB 1989, pp 185–193

Mendelzon AO, Wood PT (1995) Finding regular simple paths in graph databases. SIAM J Comput 24(6):1235–1258

Meng X, Ma ZM, Yan L (2008) Providing flexible queries over web databases. In: Knowledge-based intelligent information and engineering systems, pp 601–606

Mishra C, Koudas N (2009) Interactive query refinement. In: Proceedings of EDBT 2009, pp 862–873

Munoz S, Pérez J, Gutierrez C (2007) Minimal deductive systems for RDF. In: Proceedings of ESWC 2007, pp 53–67

Na S, Park S (2005) A process of fuzzy query on new fuzzy object oriented data model. In: Proceedings of DEXA 2005, pp 500–509

Pérez J, Arenas M, Gutierrez C (2006) Semantics and complexity of SPARQL. In: Proceedings of ISWC 2006, pp 30–43

Pérez J, Arenas M, Gutierrez C (2008) nSPARQL: a navigational language for RDF. In: Proceedings of ISWC 2008, pp 66–81

Poulovassilis A, Wood PT (2010) Combining approximation and relaxation in semantic web path queries. In: Proceedings of ISWC 2010, pp 631–646

Poulovassilis A, Selmer P, Wood PT (2012) Flexible querying of lifelong learner metadata. IEEE Trans Learn Technol 5(2):117–129

Poulovassilis A, Gutierrez-Santos S, Mavrikis M (2015) Graph-based modelling of students' interaction data from exploratory learning environments. In: Proceedings of GEDM 2015 (at Educational Data Mining 2015), pp 46–51

Poulovassilis A, Selmer P, Wood PT (2016) Approximation and relaxation of semantic web path queries. J Web Semant 40:1–21

Reddy BRK, Kumar PS (2010) Efficient approximate SPARQL querying of web of linked data. In: Proceedings of URSW 2010, pp 37–48

Sakr S, Elnikety S, He Y (2012) G-SPARQL: a hybrid engine for querying large attributed graphs. In: Proceedings of CIKM 2012, pp 335–344

Sarma D et al (2008) Bootstrapping pay-as-you-go data integration systems. In: Proceedings of SIGMOD 2008, pp 861–874

Sassi M, Tlili O, Ounelli H (2012) Approximate query processing for database flexible querying with aggregates. Trans Large-Scale Data- Knowl Centered Syst V:1–27

Selmer P (2016) Flexible querying of graph-structured data. PhD thesis, Birkbeck, University of London

Selmer P, Poulovassilis A, Wood PT (2015) Implementing flexible operators for regular path queries. In: Proceedings of GraphQ 2015 (EDBT/ICDT Workshops), pp 149–156

Siepen J et al (2008) ISPIDER Central: an integrated database web-server for proteomics. Nucleic Acids Res (Web-Server-Issue) 36:485–490

Suthers D (2015) From contingencies to network-level phenomena: multilevel analysis of activity and actors in heterogeneous networked learning environments. In: Proceedings of LAK 2015, pp 368–377

Theobald M, Schenkel R, Weikum G (2005) An efficient and versatile query engine for TopX search. In: Proceedings of VLDB 2005, pp 625–636

Vanhatalo J, Völzer H, Leymann F, Moser S (2008) Automatic workflow graph refactoring and completion. In: Proceedings of ICSOC 2008. Springer, Berlin, pp 100–115

Wood PT (2012) Query languages for graph databases. ACM SIGMOD Rec 41(1):50–60

Wu B, Ye Q, Yang S, Wang B (2009) Group CRM: a new telecom CRM framework from social network perspective. In: Proceedings of 1st ACM international workshop on complex networks meet information and knowledge management (CNIKM'09), pp 3–10

Wu Y, Yan X, Yang S (2013) Ontology-based subgraph querying. In: Proceedings of ICDE 2013, pp 697–708

Yakovets N, Godfrey P, Gryz J (2015) Towards query optimization for SPARQL property paths. arXiv preprint arXiv:150408262

Yang S, Wu Y, Sun H, Yan X (2014) Schemaless and structureless graph querying. Proc VLDB Endowment 7(7):565–576

Zhang S, Yang J, Jin W (2010) SAPPER: subgraph indexing and approximate matching in large graphs. PVLDB 3(1):1185–1194

Zhou X, Gaugaz J, Balke WT, Nejdl W (2007) Query relaxation using malleable schemas. In: Proceedings of ACM SIGMOD 2007, pp 545–556

Zhu L, Ng WK, Cheng J (2011) Structure and attribute index for approximate graph matching in large graphs. Inf Syst 36(6):958–972

Zou L, Mo J, Chen L, Ozsu MT, Zhao D (2011) gStore: answering SPARQL queries via subgraph matching. PVLDB 4(8):482–493

Chapter 5
Parallel Processing of Graphs

Bin Shao and Yatao Li

Abstract Graphs play an indispensable role in a wide range of application domains. Graph processing at scale, however, is facing challenges at all levels, ranging from system architectures to programming models. In this chapter, we review the challenges of parallel processing of large graphs, representative graph processing systems, general principles of designing large graph processing systems, and various graph computation paradigms. Graph processing covers a wide range of topics and graphs can be represented in different forms. Different graph representations lead to different computation paradigms and system architectures. From the perspective of graph representation, this chapter also briefly introduces a few alternative forms of graph representation besides adjacency list.

5.1 Overview

Graphs are important to many applications. However, large-scale graph processing is facing challenges at all levels, ranging from system architectures to programming models. There are a large variety of graph applications. We can roughly classify the graph applications into two categories: online query processing, which is usually optimized for low latency; and offline graph analytics, which is usually optimized for high throughput. For instance, deciding instantly whether there is a path between two given people in a social network belongs to the first category, while calculating PageRank for a web graph belongs to the second.

Let us start with a real-life knowledge graph query example. Figure 5.1 gives a real-life relation search example on a big knowledge graph. In a knowledge graph, queries that find the paths linking a set of given graph nodes usually give the relations between these entities. In this example, we find the relations between entities *Tom Cruise*, *Katie Holmes*, *Mimi Rogers*, and *Nicole Kidman*.

B. Shao (✉) · Y. Li
Microsoft Research Asia, Beijing, China
e-mail: binshao@microsoft.com; yatli@microsoft.com

© Springer International Publishing AG, part of Springer Nature 2018
G. Fletcher et al. (eds.), *Graph Data Management*, Data-Centric Systems
and Applications, https://doi.org/10.1007/978-3-319-96193-4_5

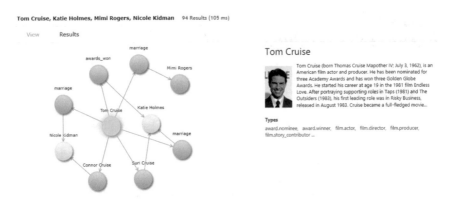

Fig. 5.1 Relation search on a knowledge graph

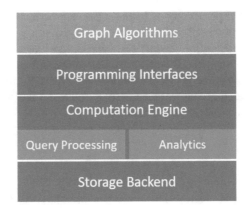

Fig. 5.2 A general graph processing system stack

Many sophisticated real-world applications highly rely on the interplay between offline graph analytics and online query processing. Given two nodes of a graph, the "distance oracle" algorithm designed by Qi et al. (2014) estimates the shortest distance between two given nodes; it is an online algorithm. However, to estimate the distances, the algorithm relies on "landmark" nodes in the graph and an optimal set of landmark nodes are discovered using an offline analytics algorithm.

Generally speaking, the system stack of a graph processing system consists of all or some of the layers shown in Fig. 5.2. At the top, graph algorithms manipulate graphs via programming interfaces provided by a graph processing system. Between the programming interfaces and the storage backend, there usually is a computation engine that executes the graph manipulation instructions dictated by the graph algorithms through programming interfaces.

At the bottom, the storage backend hosts the graph data using a certain graph representation, either in a single machine or over multiple distributed machines. Storage backends have important system design implications. The storage layer largely determines the system optimization goal, as discussed in Kaoudi and Manolescu (2015). For example, systems including SHARD by Rohloff and Schantz (2011), HadoopRDF by Husain et al. (2011), RAPID by Ravindra et al. (2011), and EAGRE by Zhang et al. (2013) use a distributed file system as their storage backends. The systems that directly use a file system as storage backend are usually optimized for throughput due to the relatively high data retrieval latency. In contrast, systems such as H2RDF by Papailiou et al. (2012), AMADA by Aranda-Andújar et al. (2012), and Trinity.RDF by Zeng et al. (2013) are optimized for better response time via the fast random data access capability provided by their key-value store backends.

In this section, we first introduce the notation and discuss why it is difficult to process large graphs. Then, we present some general principles of designing large-scale graph processing systems after a brief survey of some representative graph processing systems.

5.1.1   Notation

Let us introduce the terminology and notation that will be used throughout this chapter. A graph may refer to a topology-only mathematical concept as defined in Bollobás (1998) or a data set. In the former sense, a graph is a pair of finite sets (V, E) such that the set of edges E is a subset of the set $V \times V$. If each pair of vertices are ordered, we call G a directed graph; otherwise, we call it an undirected graph.

In what follows when we represent a data set as a graph, especially when there are data associated with the vertices, we refer to the vertices as *graph nodes* or *nodes*. Correspondingly, we call adjacent vertices *neighboring nodes*. If the data set only contains graph topology or if we only want to emphasize its graph topology, we call them vertices.

There are two common ways of representing and storing a graph: adjacency list and adjacency matrix. The way of representing a graph determines the way we can process the graph. As most graph query processing and analytics algorithms highly rely on the operator that gets adjacent vertices of a given vertex, adjacency list is usually preferred way of representing a graph especially when the graph is large. If we use the adjacency matrix representation, we need to scan a whole adjacency matrix row to access the adjacent vertices of given vertex. For a graph with billions of vertices, the costs of scanning matrix rows will be prohibitive. In this chapter, we assume graphs are represented and stored as adjacency lists unless otherwise stated.

5.1.2   Challenges of Large Graph Processing

It is difficult to process large graphs mostly because they have a large number of encoded relations. We summarize the challenges of large graph processing as:

(1) the complex nature of graph; (2) the diversity of graphs; (3) the diversity of graph computations; (4) the scale of graph size.

5.1.2.1 The Complex Nature of Graph

Graphs are inherently complex. The contemporary computer architectures are good at processing linear and simple hierarchical data structures, such as *Lists*, *Stacks*, or *Trees*. When the data scale goes large, the *divide and conquer* computation paradigm still works well for these data structures, even if the data is partitioned over distributed machines.

However, when we are handling graphs, especially large graphs, the situation is changed. Andrew Lumsadine and Douglas Gregor (2007) summarize the characteristics of parallel graph processing as *data-driven computations*, *unstructured problems*, *poor locality*, and *high data access to computation ratio*. The implication is twofold: (1) From the perspective of data access, a graph node's neighboring nodes cannot be accessed without "jumping" in the storage no matter how we represent a graph. In other words, a large amount of random data accesses are required during graph processing. Many modern program optimization techniques rely on data locality and data reuse. Unfortunately, the random data access nature of graph breaks the premise. This usually causes poor system performance as the CPU cache is not effective for most of the time. (2) From the perspective of program structure, it is difficult to extract parallelism because of the unstructured nature of graphs. Partitioning large graphs itself is an NP-hard problem as shown by Garey et al. (1974); this makes it hard to get an efficient *divide and conquer* solution for many large graph processing tasks.

5.1.2.2 The Diversity of Graphs

There are many kinds of graphs, such as scale-free graphs, graphs with community structures, and small-world graphs. A scale-free graph is a graph whose degrees follow a power-law distribution. For graphs with community structure, the graph nodes can easily be grouped into sets of nodes such that each set of nodes are densely connected. For small-world graphs, most nodes can be reached from other nodes by a small number of hops. The performance of graph algorithms may vary a lot on different kinds of graphs.

5.1.2.3 The Diversity of Graph Computations

Furthermore, there are a large variety of graph computations. As discussed earlier, graph computations can be roughly classified into two categories: online query processing and offline graph analytics. Besides common graph query processing and analytics tasks, there are other useful graph operations such as graph generation, graph visualization, and interactive exploration. It is challenging to design a system that can support all these operations on top of the same infrastructure.

5.1.2.4 The Scale of Graph Size

Last but not least, the scale of graph size does matter. Graphs with billions of nodes are common now, for example, the Facebook social network has more than two billion monthly active users.[1] The World Wide Web has more than one trillion unique links. The De brujin graph for genes even has more than one trillion nodes and at least eight trillion edges. The scale of graph size makes many classic graph algorithms from textbooks ineffective.

5.1.3 Representative Graph Processing Systems

Recent years have witnessed an explosive growth of graph processing systems as shown by Aggarwal and Wang (2010). However, many graph algorithms are ad hoc in the sense that each of them assumes that the underlying graph data is organized in a certain way to maximize its performance. In other words, there is not a standard or de facto graph system on which graph algorithms are developed and optimized. In response to this situation, a number of graph systems have been proposed. Some representative systems are summarized in Table 5.1.

Neo4j[2] focuses on supporting online transaction processing (OLTP) of graphs. Neo4j is like a regular database system, with a more expressive and powerful data model. Its computation model does not handle graphs that are partitioned over multiple machines. For large graphs that cannot be stored in the main memory, disk random access becomes the performance bottleneck.

From the perspective of online graph query processing, a few distributed in-memory systems have been designed to meet the challenges faced by disk-based single-machine systems. Representative systems include Trinity by Shao et al. (2013) and Horton by Sarwat et al. (2013). These systems leverage RAM to speed up random data accesses and use a distributed computation engine to process graph queries in parallel.

On the other end of the spectrum are MapReduce by Dean and Ghemawat (2008), PEGASUS by Kang et al. (2009), Pregel by Malewicz et al. (2010), Giraph, GraphLab by Low et al. (2012), GraphChi by Kyrola et al. (2012), and GraphX by Gonzalez et al. (2014). These systems are usually not optimized for online query processing. Instead, they are optimized for high-throughput analytics on large graphs partitioned over many distributed machines.

---

[1] http://newsroom.fb.com/company-info/.
[2] http://neo4j.com/.

Table 5.1 Some representative graph processing systems (*SB* means the feature depends on its storage backend)

	Native graphs	Online query	Data sharding	In-memory storage	Transaction support
Neo4j	Yes	Yes	No	No	Yes
Trinity	Yes	Yes	Yes	Yes	Atomicity
Horton	Yes	Yes	Yes	Yes	No
FlockDB[a]	No	Yes	Yes	No	Yes
TinkerGraph[b]	Yes	Yes	No	Yes	No
InfiniteGraph[c]	Yes	Yes	Yes	No	Yes
Cayley[d]	Yes	Yes	SB	SB	Yes
Titan[e]	Yes	Yes	SB	SB	Yes
MapReduce	No	No	Yes	No	No
PEGASUS	No	No	Yes	No	No
Pregel	No	No	Yes	No	No
Giraph[f]	No	No	Yes	No	No
GraphLab	No	No	Yes	No	No
GraphChi	No	No	No	No	No
GraphX	No	No	Yes	No	No

[a]https://github.com/twitter/flockdb
[b]https://github.com/tinkerpop/blueprints/wiki/TinkerGraph
[c]http://www.objectivity.com/products/infinitegraph/
[d]https://github.com/google/cayley
[e]http://thinkaurelius.github.io/titan/
[f]http://giraph.apache.org/

MapReduce-based graph processing depends heavily on inter-processor bandwidth as graph structures are sent over the network iteration after iteration. Pregel and its follow-up systems mitigate this problem by passing computation results instead of graph structures between processors. In Pregel, analytics are expressed using a vertex-centric computation paradigm. Although some well-known graph algorithms such as PageRank and shortest path discovery can be expressed using vertex-centric computation paradigm easily; there are many sophisticated graph computations that cannot be expressed in a succinct and elegant way.

The systems listed in Table 5.1 are compared from five aspects. First, does the graph exist in its native form? When a graph is in its native form, graph algorithms can be expressed in standard, natural ways as discussed by Cohen (2009). Second, does the system support low-latency query processing? Third, does the system support data sharding and distributed parallel graph processing? Fourth, does the system use RAM as the main storage? Fifth, does the system support transactions?

5.2   General Design Principles

We have reviewed a few representative graph processing systems. In this section, we discuss a few general principles of designing a general-purpose, real-time graph processing system.

5.2.1   Addressing the Grand Random Data Access Challenge

As discussed earlier, a graph node's neighboring nodes' cannot be accessed without "jumping" in the storage no matter how we represent a graph. A lot of random accesses on hard disks lead to performance bottlenecks. It is important to keep graphs memory-resident for efficient graph computations, especially for real-time online query processing. In order to create a general-purpose graph processing system that supports both low-latency online query processing and high-throughput offline analytics, the grand challenge of random accesses must be well addressed at the data access layer.

Despite the great progress made in disk technology, it still cannot provide the level of efficient random access required for graph processing. DRAM (dynamic random-access memory) is still the only promising storage medium that can provide a satisfactory level of random access performance with acceptable costs. On the other hand, in-memory approaches usually cannot scale to very large graphs due to the capacity limit of a single machine. We argue that distributed RAM storage is a promising approach to efficient large graph processing.

By addressing the random data access challenge, we can design systems that efficiently support both online graph query processing and offline graph analytics instead of optimizing the systems for certain graph computations. For online queries, it is particularly effective to keep graphs in-memory, as most online query processing algorithms, such as BFS, DFS, and subgraph matching, require a certain degree of graph exploration. On the other hand, offline graph computations are usually conducted in an iterative, batch manner. For iterative computations, keeping data in RAM can boost the performance by an order of magnitude due to the reuse of intermediate results as discussed by Zaharia et al. (2010).

5.2.2   Avoiding Prohibitive Indexes

For offline graph analytics, partitioning the computation task well (if possible) is, to some extent, the silver bullet. As long as we can find an efficient way to partition the graph data, we basically have a solution to efficient offline graph processing. Due to the random data access challenge, general-purpose, efficient disk-based solutions usually do not exist. But under certain constraints, offline graph analytics tasks can have efficient disk-based "divide and conquer" solutions. A good example is the GraphChi system proposed by Kyrola et al. (2012). If a computational problem can be well-partitioned, then the subproblems can be loaded in the main memory and processed in-memory one at a time. However, as discussed by Lumsdaine et al. (2007), many graph problems are inherently irregular and computation partitioning is hard, especially for online queries.

Compared with offline analytics, online queries are much harder to handle due to the following two reasons. First, online queries are sensitive to network latency. It is harder to reduce communication latency than to increase throughput by adding more machines. On the one hand, adding more machines can reduce each machine's workload; on the other hand, having more machines incurs higher communication costs. Second, it is generally difficult to predict the data access patterns of a graph query, thus it is hard to optimize the execution by leveraging I/O optimization techniques such as prefetching.

Many graph computations are I/O intensive; data accesses usually dominate the graph computation costs. The performance of processing a graph query depends on how fast we randomly access the graph. A traditional way of speeding up random data access is to use indexes. Graph indexes are widely employed to speed up online query processing, either by precomputing and materializing the results of common query patterns or by storing redundant information. To capture the structural information of graphs, graph indexes usually require super-linear indexing time and super-linear storage space. For large graphs, for example, graphs with billions of nodes, the super-linear complexity means infeasibility. We will show in the following section that index-free graph processing is a possible and efficient approach to many real-time graph query processing tasks.

5.2.3   Supporting Fine-Grained One-Sided Communications

Most graph computations are data-driven and the communication costs typically contribute a large portion to the overall system costs. Overlapping computations well with the underlying communication is the key to high performance.

MPI is the de facto standard for message passing programming in high-performance computing and pairwise two-sided send/receive communication is the major paradigm provided by MPI.[3] The communication progress is dictated by explicitly invoked MPI primitive calls as shown by Majumder and Rixner (2004). Nearly all modern network communication infrastructures provide asynchronous network events. MPI communication paradigm incurs unnecessary latency because it responds to network events only during *send* and *receive* primitive invocations.

In contrast, active messages introduced by von Eicken et al. (1992) is a communication architecture that can well overlap computation and communication.

---

[3]Even one-sided primitives are included starting from MPI-2 standard, their usage is still limited.

The communication architecture is desirable for data-driven graph computation, especially for online graph query processing, which is sensitive to network latency.

When we use active messages, a user-space message handler will be invoked upon the arrival of a message. The message handler is pointed by a handler index encoded in the message. Let us use a simple example to illustrate the difference between the two-sided communication paradigm and active messages. Suppose we want to send some messages from one machine to another according to the output values of a random number generator. Using active messages, we can check the random values on the sender side and invoke a *send* operation only if the value matches certain sending condition. Using the pairwise two-sided communication paradigm, we need to invoke as many send/receive calls as the number of generated random values and perform the value checkings on the receiver side.

5.3 Online Query Processing

In this section, we review two online query processing techniques specially designed for distributed large graphs: asynchronous fanout search and index-free query processing.

5.3.1 *Asynchronous Fanout Search*

Most graph algorithms require a certain kind of graph exploration; breadth-first search (BFS) and depth-first search (DFS) are among the commonest. Here, we use people search in a social network as an example to demonstrate an efficient graph exploration technique called *asynchronous fanout search*. The problem is the following: given a user of a social network, find the people whose first name is "David" among his friends, his friends' friends, and his friends' friends' friends.

It is unlikely that we can index the social network to solve the "David" problem. One option is to index the neighboring nodes for each user, so that given any user, we can use the index to check if there is "David" within his or her 3-hop neighborhood.

Algorithm 1: Asynchronous Fanout search

Require: v (a given graph node)
Ensure: all "Davids" within the 3-hop neighboring nodes
 1: $N \leftarrow$ the ids of v's neighboring nodes
 2: $k \leftarrow$ (the number of machines)
 3: $hop \leftarrow 1$
 4: Partition N into k parts: $N = N_1 \cup \cdots \cup N_k$
 5: **parallel-foreach** N_i in N
 6:      async_send message (N_i, hop) to machine i

Algorithm 2: On receiving message (N_i, hop)

1: S_i ← the graph nodes with ids in N_i
2: check if there are "Davids" in S_i
3: **if** $hop < 3$ **then**
4:      N' ← ids of the neighboring nodes of the nodes in S_i
5:      Partition N' into k parts: $N' = N'_1 \cup \cdots \cup N'_k$
6:      **parallel-foreach** N'_i in N'
7:          async_send message $(N'_i, hop + 1)$ to machine i

---

However, the size and the update cost of such an index are prohibitive for a large graph. The second option is to create an index to answer 3-hop reachability queries for any two nodes. This is infeasible either because "David" is a popular name and we cannot check every "David" in the social network to see whether he is within 3 hops to the current user.

We can tackle the "David" problem by leveraging fast memory-based graph explorations. The algorithm simply sends asynchronous "fan-out search" requests recursively to remote machines as shown by Algorithms 1 and 2. Specifically, it partitions v's neighboring nodes into k parts $N_1, N_2, \ldots, N_k$ (line 4 of Algorithm 1), where k is the total number of machines. Then, the "fan-out" search is performed by sending message (N_i, hop) (line 6 of Algorithm 1) to all machines in parallel. On receiving the search requests, machine i searches for "David" in its local data storage (line 2 of Algorithm 2) and sends out the next-hop "fan-out" search requests $(N'_i, hop + 1)$ (line 7 of Algorithm 2).

This simple fanout search works well for randomly partitioned large distributed graphs. As demonstrated by Shao et al. (2013), for a Facebook-like graph, exploring the entire 3-hop neighborhood of any given node in the graph takes less than 100 ms on average.

5.3.2 Index-Free Query Processing

It is usually harder to optimize online query processing because of its limited response time budget. We use the subgraph matching problem as an example to introduce an efficient online query processing paradigm for distributed large graphs.

Subgraph matching is the basic graph operation underpinning many graph applications. Graph $G' = (V', E')$ is a subgraph of $G = (V, E)$ if $V' \subset V$ and $E' \subset E$. Graph $G'' = (V'', E'')$ is isomorphic to $G' = (V', E')$ if there is a bijection $f : V' \rightarrow V''$ such that $xy \in E'$ iff $f(x)f(y) \in E''$. For a given data graph G and a query graph G', subgraph matching is to retrieve all the subgraphs of G that are isomorphic to the query graph.

Canonical subgraph matching algorithms are usually conducted in the following three steps:

1. Break the data graph into basic units such as edges, paths, or frequent subgraphs.
2. Build indexes for the basic units.
3. Decompose a query into multiple basic unit queries, do subgraph matching for the unit queries, and join their results.

It is much more costly to index graph structure than to index a relational table. For instance, 2-hop reachability indexes usually require $O(n^4)$ construction time. Depending on the structure of the basic unit, the space costs vary. In many cases, they are super-linear. Furthermore, multiway joins are costly too, especially when the data is disk-resident.

To demonstrate the infeasibility of index-based solutions for large graphs, let us review a survey on subgraph matching made by Sun et al. (2012), as shown in Tables 5.2 and 5.3.

Table 5.2 shows the index costs of a few representative subgraph matching algorithms proposed by Ullmann (1976), Cordella et al. (2004), Neumann and Weikum (2010), Atre et al. (2010), Holder et al. (1994), Zhu et al. (2011), Cheng et al. (2008), Zou et al. (2009), He and Singh (2008), Zhao and Han (2010), Zhang et al. (2009), and Sun et al. (2012). To illustrate what the costs listed in Table 5.2 imply for a large graph, Table 5.3 shows their estimated index construction costs and query time for a Facebook-like social network. Although RDF-3X and BitMat have linear indexing complexity, they take more than 20 days to index a Facebook-like large graph, let alone those super-linear indexes. The evident conclusion is that the costly graph indexes are infeasible for large graphs.

To avoid building sophisticated indexes, the STwig method proposed by Sun et al. (2012) and the Trinity.RDF system proposed by Zeng et al. (2013) process subgraph matching queries without using structural graph indexes. This ensures scalability;

Table 5.2 Survey on subgraph matching algorithms by Sun et al. (2012)

Algorithms	Index size	Index time	Update cost	Graph size in experiments
Ullmann, VF2	–	–	–	4484
RDF-3X	$O(m)$	$O(m)$	$O(d)$	33M
BitMat	$O(m)$	$O(m)$	$O(m)$	361M
Subdue	–	Exponential	$O(m)$	10K
SpiderMine	–	Exponential	$O(m)$	40K
R-join	$O(nm^{1/2})$	$O(n^4)$	$O(n)$	1M
Distance-join	$O(nm^{1/2})$	$O(n^4)$	$O(n)$	387K
GraphQL	$O(m + nd^r)$	$O(m + nd^r)$	$O(d^r)$	320K
Zhao	$O(nd^r)$	$O(nd^r)$	$O(d^L)$	2M
GADDI	$O(nd^L)$	$O(nd^L)$	$O(d^L)$	10K
STwig	$O(n)$	$O(n)$	$O(1)$	1B

Table 5.3 Index costs and query time of subgraph matching algorithms for a Facebook-like graph

Algorithms	Index size	Index time	Query time
Ullmann, VF2	–	–	>1000
RDF-3X	1T	>20 days	>48
BitMat	2.4T	>20 days	>269
Subdue	–	>67 years	–
SpiderMine	–	>3 years	–
R-join	>175T	>10^{15} years	>200
Distance-join	>175T	>10^{15} years	>4000
GraphQL	>13T(r=2)	>600 years	>2000
Zhao	>12T(r=2)	>600 years	>600
GADDI	>2 × 10^5T (L=4)	>4 × 10^5 years	>400
STwig	6G	33 s	<20

they work well for graphs with billions of nodes, which are not *indexable* in terms of both index space and index time.

To compensate for the performance loss due to the lack of structural indexes, both STwig and Trinity.RDF heavily make use of in-memory graph explorations to replace costly join operations. Given a query, they split it into a set of subqueries that can be efficiently processed via in-memory graph exploration. They only perform join operations when they are absolutely necessary and not avoidable, for example, when there is a cycle in the query graph. This dramatically reduces query processing time, which is usually dominated by join operations.

5.4 Offline Analytics

Graph analytics jobs perform a global computation against a graph. Many of them are conducted in an iterative manner. When the graph is large, the analytics jobs are usually conducted as offline tasks.

In this section, we review the MapReduce computation paradigm and vertex-centric computation paradigm for offline graph analytics. Then, we discuss communication optimization and a lightweight analytics technique called local sampling.

5.4.1 MapReduce Computation Paradigm

MapReduce, as elaborated by Dean and Ghemawat (2008), is a high-latency yet high-throughput data processing platform that is optimized for offline analytics for large partitioned data sets. MapReduce is a very successful programming model for big data processing. However, when used for processing graphs, it suffers from the following problems: First, it is very hard to support real-time online queries. Second, the data model of MapReduce cannot model graphs natively and graph algorithms cannot be expressed intuitively. Third, MapReduce highly relies on data partitioning; however, it is inherently hard to partition graphs.

It is possible to run a graph processing job efficiently on a MapReduce platform if the graph could be well-partitioned. Some well-designed graph algorithms implemented in MapReduce are given by Qin et al. (2014). The computation parallelism a MapReduce system can achieve depends on how well the data can be partitioned. Unfortunately, the partitioning task of a large graph can be very costly, as elaborated by Wang et al. (2014).

5.4.2 Vertex-Centric Computation Paradigm

The vertex-centric graph computation paradigm, which was first advocated by Malewicz et al. (2010), provides a vertex-centric computational abstraction over the BSP model proposed by Valiant (1990). A computation task is expressed in multiple iterative super-steps and each vertex acts as an independent agent. During each super-step, each agent performs some computations, independent of each other. It then waits for all other agents to finish their computations before the next super-step begins.

Compared with MapReduce, Pregel exploits finer-grained parallelism at the vertex level. Moreover, Pregel does not move graph partitions over the network; only messages among graph vertices are passed at the end of each iteration. This greatly reduces the network traffic.

Many follow-up works, such as GraphLab by Low et al. (2012), PowerGraph by Gonzalez et al. (2012), Trinity by Shao et al. (2013), and GraphChi by Kyrola et al. (2012), support the vertex-centric computation paradigm for offline graph analytics. Among these systems, GraphChi is specially worth mentioning as it well addresses the "divide and conquer" problem for graph computation under certain constraints. GraphChi can perform efficient disk-based graph computation as long as the computation could be expressed as an asynchronous vertex-centric algorithm. An asynchronous algorithm is one where a vertex can perform a computational task based solely on the partially updated information from its incoming links. This assumption, on the one hand, frees the need of passing messages from the current vertex to all its outgoing links so that it can perform the graph computations block by block. On the other hand, it inherently cannot efficiently support traversal-based graph computations and synchronous graph computations because it cannot freely access the outgoing links of a vertex.

Although quite a few graph computation tasks, including Single Source Shortest Paths, PageRank, and their variants, can be expressed elegantly using the vertex-centric computation paradigm, there are many that cannot be elegantly and intuitively expressed using the vertex-centric paradigm, for example, multilevel graph partitioning.

5.4.3 Communication Optimization

Communication optimization is very important for distributed graph computation. Although a graph is distributed over multiple machines, from the point view of a local machine, vertices of the graph are in two categories: vertices on the local machine, and vertices on any of the remote machines. Figure 5.3 shows a local machine's bipartite view of the entire graph.

Let us take the vertex-centric computation as an example. One naive approach is to run jobs on local vertices without preparing any messages in advance. When a local vertex is scheduled to run a job, we obtain remote messages for the vertex and run the job immediately after they arrive. Since the system usually does not have space to hold all messages, we discard messages after they are used. For instance, in Fig. 5.3, in order to run the job on vertex x, we need messages from vertices u, v, and others. Later on, when y is scheduled to run, we need messages from u and v again. This means a single message needs to be delivered multiple times, which is unacceptable in an environment where network capacity is an extremely valuable resource.

Some graph processing systems, such as the ones built using Parallel Boost Graph Library (PBGL), use *ghost nodes* (local replicas of remote nodes) for message passing as elaborated by Gregor and Lumsdaine (2005). This mechanism works well for well-partitioned graphs. However, it is difficult to create partitions of even size while minimizing the number of edge cuts. A great memory overhead would be incurred for a large graph if it is not well-partitioned. To illustrate the

Fig. 5.3 Bipartite view on a local machine

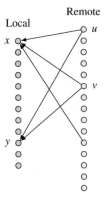

Fig. 5.4 Breadth-first search using PBGL

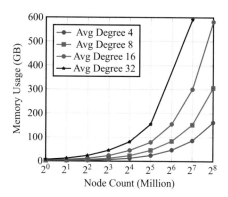

memory overhead, Fig. 5.4 shows the memory usage for graphs with 1 million to 256 million vertices. It takes nearly 600 GB main memory for the 256-million-node graph when the average degree is 16.

The messages are usually too big to be RAM-resident. We will have a great performance penalty, if we buffer the messages on the disk and perform random accesses on the disk. To address this issue, we can cache the messages in a smarter way. For example, on each machine, we can differentiate remote vertices into two categories. The first category contains hub vertices, that is, vertices having a large degree and connecting to many local vertices. The second category contains the remaining vertices. We buffer messages from the vertices in the first category for the entire duration of one computation iteration. For a scale-free graph, for example, one generated by degree distribution $P(k) \sim ck^{-\gamma}$ with $c = 1.16$ and $\gamma = 2.16$, 20% hub vertices are sending messages to 80% of vertices. Even if we only buffer messages from 10% hub vertices, we have addressed 72.8% of the message needs.

5.4.4 Local Sampling

In a distributed graph processing system, a large graph is partitioned and stored on a number of distributed machines. This leads to the following question: Can we perform graph computations locally on *each machine* and then aggregate their results to derive the final result for the entire graph? Furthermore, can we use probabilistic inferences to derive the result for the entire graph from the result on *a single machine*? This paradigm has the potential to overcome the network communication bottleneck, as it minimizes or even totally eliminates the network communication. The answers to these questions are positive. If a graph is partitioned over ten machines, each machine has full information about 10% of the vertices and 10% of the edges. Also, the edges link to a large amount of the remaining 90% of the vertices. Thus, each machine actually contains a great deal of information about the entire graph.

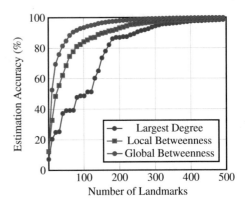

Fig. 5.5 Effectiveness of local sampling in distance oracle

The distance oracle proposed by Qi et al. (2014) demonstrated this graph computation paradigm. Distance oracle finds landmark vertices and uses them to estimate the shortest distances between any two vertices in a large graph. Figure 5.5 shows the effectiveness of three methods for picking landmark vertices. Here, the X axis shows the number of used landmark vertices and the Y axis shows estimation accuracy. The best approach is to use vertices that have the highest global betweenness and the worst approach is to simply use vertices that have the largest degree. The distance oracle approach uses the vertices that have the highest betweenness computed locally. Its accuracy is close to the best approach and its computation costs are dramatically less than that of calculating the highest global betweenness.

5.5 Alternative Graph Representations

In the previous sections, we assume the graph is modeled and stored in the adjacency list form. For a certain task, transforming a graph to other representation forms can help tackle the problem. This section covers three of them: matrix arithmetic, graph embedding, and matroids.

5.5.1 Matrix Arithmetic

A representative system is Pegasus by Kang et al. (2009). Pegasus is an open source large graph mining system. The key idea of Pegasus is to convert graph mining operations into iterative matrix–vector multiplications.

Pegasus uses an n by n matrix m and a vector v of size n to represent graphs. Pegasus defines an operation called *Generalized Iterated Matrix–Vector*

Multiplication (GIM-V).

$$M \times v = v', \text{ where } v'_i = \Sigma_{j=1}^{n} m_{i,j} \times v_j$$

Based on it, three primitive graph mining operations are defined. Graph mining problems are solved by customizing the following three operations:

- *combine2*$(m_{i,j}, v_j)$: multiply $m_{i,j}$ and v_j;
- *combineAll*$_i(x_1, \ldots, x_n)$: sum the multiplication results from *combine2* for node i;
- *assign*(v_i, v_{new}): decide how to update v_i with v_{new}.

Many graph mining algorithms, including PageRank, random walk, and connected component, can be expressed elegantly using these three customized primitive operations.

5.5.2 Graph Embedding

Some graph problems can be easily solved after we embed a graph into a high-dimensional space as illustrated by Zhao et al. (2010, 2011) and Qi et al. (2014). This approach is particularly useful for estimating the distances between graph nodes.

Let us use an example given by Zhao et al. (2011) to illustrate the main idea of high-dimensional graph embedding. To compute the distance between two given graph vertices, we can embed a graph into a geometric space so that the distances in the space preserve the shortest distances in the graph. In this way, we can immediately give an approximate shortest distance between two vertices by calculating the Euclidean distance between their coordinates in the high-dimensional geometric space.

5.5.3 Matroids

As illustrated by Oxley (1992), any undirected graph can be represented by a binary matrix that in turn can produce a graphic matroid. Matroids usually use an *"edge-centric"* graph representation. As elaborated by Truemper (1998), instead of representing a graph as (V, E), we consider a graph as a set E of edges and consider graph nodes as certain subsets of E. For example, a graph $(V, E) = (\{a, b, c\}, \{e_1 = ab, e_2 = bc, e_3 = ca\})$ can be represented as follows: $E = \{e_1, e_2, e_3\}$ and $V = \{a = \{e_1, e_3\}, b = \{e_1, e_2\}, c = \{e_2, e_3\}\}$.

Matroids provide a new angle of looking at graphs with a powerful set of tools for solving many graph problems. Even though the interplay between graphs and matroids has been proven fruitful, as elaborated by Oxley (2001), how matroids can be leveraged to help design graph processing systems is still an open problem.

5.6 Summary

The proliferation of large graph applications demands efficient graph processing systems. Parallel graph processing is an active research area. This chapter tried to shed some light on parallel large graph processing from a pragmatic point of view. We discussed the challenges and general principles of designing a general-purpose, large-scale graph processing system. After surveying a few representative systems, we reviewed a few important graph computation paradigms for online query processing and offline analytics. Different graph representations lead to different graph computation paradigms; each of them may be suitable for solving a certain range of problems. At the end of this chapter, we briefly explored a few alternative graph representation forms and their applications.

References

Aggarwal CC, Wang H (eds) (2010) Managing and mining graph data. Advances in database systems, vol 40. Springer, Berlin

Aranda-Andújar A, Bugiotti F, Camacho-Rodríguez J, Colazzo D, Goasdoué F, Kaoudi Z, Manolescu I (2012) Amada: web data repositories in the amazon cloud. In: Proceedings of the 21st ACM international conference on information and knowledge management, CIKM '12. ACM, New York, pp 2749–2751

Atre M, Chaoji V, Zaki MJ, Hendler JA (2010) Matrix "bit" loaded: a scalable lightweight join query processor for RDF data. In: WWW, pp 41–50

Bollobás B (1998) Modern graph theory. Graduate texts in mathematics, Springer, Berlin

Cheng J, Yu JX, Ding B, Yu PS, Wang H (2008) Fast graph pattern matching. In: ICDE, pp 913–922

Cohen J (2009) Graph twiddling in a mapreduce world. In: Computing in science & engineering, pp 29–41

Cordella LP, Foggia P, Sansone C, Vento M (2004) A (sub)graph isomorphism algorithm for matching large graphs. IEEE Trans Pattern Anal Mach Intell 26(10):1367–1372

Dean J, Ghemawat S (2008) Mapreduce: simplified data processing on large clusters. Commun ACM 51:107–113

Garey MR, Johnson DS, Stockmeyer L (1974) Some simplified np-complete problems. In: Proceedings of the sixth annual ACM symposium on theory of computing, STOC '74. ACM, New York, pp 47–63

Gonzalez JE, Low Y, Gu H, Bickson D, Guestrin C (2012) Powergraph: distributed graph-parallel computation on natural graphs. In: OSDI, pp 17–30

Gonzalez JE, Xin RS, Dave A, Crankshaw D, Franklin MJ, Stoica I (2014) Graphx: graph processing in a distributed dataflow framework. In: Proceedings of the 11th USENIX conference on operating systems design and implementation, OSDI'14. USENIX Association, Berkeley, pp 599–613

Gregor D, Lumsdaine A (2005) The parallel BGL: a generic library for distributed graph computations. In: Parallel object-oriented scientific computing (POOSC), POOSC '05

He H, Singh AK (2008) Graphs-at-a-time: query language and access methods for graph databases. In: SIGMOD

Holder LB, Cook DJ, Djoko S (1994) Substucture discovery in the subdue system. In: KDD workshop, pp 169–180

Husain M, McGlothlin J, Masud MM, Khan L, Thuraisingham BM (2011) Heuristics-based query processing for large RDF graphs using cloud computing. IEEE Trans Knowl Data Eng 23(9):1312–1327

Kang U, Tsourakakis CE, Faloutsos C (2009) Pegasus: a peta-scale graph mining system implementation and observations. In: Proceedings of the 2009 ninth IEEE international conference on data mining, ICDM '09. IEEE Computer Society, Washington, pp 229–238

Kaoudi Z, Manolescu I (2015) RDF in the clouds: a survey. VLDB J 24(1):67–91

Kyrola A, Blelloch G, Guestrin C (2012) Graphchi: large-scale graph computation on just a pc. In: OSDI, pp 31–46

Low Y, Bickson D, Gonzalez J, Guestrin C, Kyrola A, Hellerstein JM (2012) Distributed graphlab: a framework for machine learning and data mining in the cloud. Proc VLDB Endow 5(8):716–727

Lumsdaine A, Gregor D, Hendrickson B, Berry JW (2007) Challenges in parallel graph processing. Parallel Process Lett 17(1):5–20

Majumder S, Rixner S (2004) An event-driven architecture for MPI libraries. In: Proceedings of the 2004 Los Alamos computer science institute symposium

Malewicz G, Austern MH, Bik AJ, Dehnert JC, Horn I, Leiser N, Czajkowski G (2010) Pregel: a system for large-scale graph processing. In: Proceedings of the 2010 international conference on management of data, SIGMOD '10. ACM, New York, pp 135–146

Neumann T, Weikum G (2010) The rdf-3x engine for scalable management of RDF data. VLDB J 19(1):91–113

Oxley J (1992) Matroid theory. Oxford University Press, Oxford

Oxley J (2001) On the interplay between graphs and matroids. In: Surveys in combinatorics 2001. Cambridge University Press, Cambridge

Papailiou N, Konstantinou I, Tsoumakos D, Koziris N (2012) H2rdf: adaptive query processing on RDF data in the cloud. In: Proceedings of the 21st international conference on World Wide Web, WWW '12 Companion. ACM, New York, pp 397–400

Qi Z, Xiao Y, Shao B, Wang H (2014) Distance oracle on billion node graphs. In: VLDB, VLDB Endowment

Qin L, Yu JX, Chang L, Cheng H, Zhang C, Lin X (2014) Scalable big graph processing in mapreduce. In: Proceedings of the 2014 ACM SIGMOD international conference on management of data, SIGMOD '14. ACM, New York, pp 827–838

Ravindra P, Kim H, Anyanwu K (2011) An intermediate algebra for optimizing RDF graph pattern matching on mapreduce. In: Proceedings of the 8th extended semantic web conference on the semanic web: research and applications - volume Part II, ESWC'11. Springer, Berlin, pp 46–61

Rohloff K, Schantz RE (2011) Clause-iteration with mapreduce to scalably query datagraphs in the shard graph-store. In: Proceedings of the fourth international workshop on data-intensive distributed computing, DIDC '11. ACM, New York, pp 35–44

Sarwat M, Elnikety S, He Y, Mokbel MF (2013) Horton+: a distributed system for processing declarative reachability queries over partitioned graphs. Proc VLDB Endow 6(14):1918–1929

Shao B, Wang H, Li Y (2013) Trinity: a distributed graph engine on a memory cloud. In: Proceedings of the 2013 ACM SIGMOD international conference on management of data, SIGMOD '13. ACM, New York, pp 505–516

Sun Z, Wang H, Wang H, Shao B, Li J (2012) Efficient subgraph matching on billion node graphs. Proc VLDB Endow 5(9):788–799

Truemper K (1998) Matroid decomposition. Elsevier, Amsterdam

Ullmann JR (1976) An algorithm for subgraph isomorphism. J ACM 23(1):31–42

Valiant LG (1990) A bridging model for parallel computation. Commun ACM 33:103–111 von Eicken T, Culler DE, Goldstein SC, Schauser KE (1992) Active messages: a mechanism for integrated communication and computation. In: Proceedings of the 19th annual international symposium on computer architecture, ISCA '92. ACM, New York, pp 256–266

Wang L, Xiao Y, Shao B, Wang H (2014) How to partition a billion-node graph. In: IEEE 30th international conference on data engineering, ICDE 2014, Chicago, March 31–April 4, 2014, pp 568–579

Zaharia M, Chowdhury M, Franklin MJ, Shenker S, Stoica I (2010) Spark: cluster computing with working sets. In: HotCloud'10 proceedings of the 2nd USENIX conference on hot topics in cloud computing. USENIX Association, Berkeley, 18 pp.

Zeng K, Yang J, Wang H, Shao B, Wang Z (2013) A distributed graph engine for web scale RDF data. In: VLDB, VLDB Endowment

Zhang S, Li S, Yang J (2009) Gaddi: distance index based subgraph matching in biological networks. In: EDBT

Zhang X, Chen L, Tong Y, Wang M (2013) Eagre: towards scalable I/O efficient SPARQL query evaluation on the cloud. In: Proceedings of the 2013 IEEE international conference on data engineering (ICDE 2013), ICDE '13. IEEE Computer Society, Washington, pp 565–576

Zhao P, Han J (2010) On graph query optimization in large networks. PVLDB 3(1):340–351

Zhao X, Sala A, Wilson C, Zheng H, Zhao BY (2010) Orion: shortest path estimation for large social graphs. In: WOSN'10

Zhao X, Sala A, Zheng H, Zhao BY (2011) Fast and scalable analysis of massive social graphs. CoRR

Zhu F, Qu Q, Lo D, Yan X, Han J, Yu PS (2011) Mining top-k large structural patterns in a massive network. In: VLDB

Zou L, Chen L, Özsu MT (2009) Distancejoin: pattern match query in a large graph database. PVLDB 2(1):886–897

Chapter 6
A Survey of Benchmarks for Graph-Processing Systems

Angela Bonifati, George Fletcher, Jan Hidders, and Alexandru Iosup

Abstract Benchmarking is a process that informs the public about the capabilities of systems-under-test, focuses on expected and unexpected system-bottlenecks, and promises to facilitate system tuning and new systems designs. In this chapter, we survey benchmarking approaches for graph-processing systems. First, we describe the main features of a benchmark for graph-processing systems. Then, we systematically survey across these features a diverse set of benchmarks for RDF databases, benchmarks for graph databases, benchmarks for parallel and distributed graph-processing systems, and data-only benchmarks. We trace in our survey not only the important benchmarks, but also their innovative approaches and how their core ideas evolved from previous benchmarking approaches. Last, we identify ongoing and future research directions for benchmarking initiatives.

A. Bonifati (✉)
Université Claude Bernard Lyon 1, Villeurbanne, France
e-mail: angela.bonifati@univ-lyon1.fr

G. Fletcher
Department of Mathematics and Computer Science, Eindhoven University of Technology, Eindhoven, the Netherlands
e-mail: g.h.l.fletcher@tue.nl

J. Hidders
Department of Computer Science, Vrije Universiteit Brussel, Brussels, Belgium
e-mail: jan.hidders@vub.be

A. Iosup
Vrije Universiteit Amsterdam, Amsterdam, the Netherlands

Delft University of Technology, Delft, the Netherlands
e-mail: A.Iosup@vu.nl

© Springer International Publishing AG, part of Springer Nature 2018
G. Fletcher et al. (eds.), *Graph Data Management*, Data-Centric Systems
and Applications, https://doi.org/10.1007/978-3-319-96193-4_6

6.1   Introduction

In computer science and computer engineering, the role of a benchmark is to provide
a standardized test that allows us to evaluate and compare the performance of certain
computer systems. Such tests are important in different ways. One is that it allows
the prospective users of these systems to decide on the basis of relatively objective
measurements which one is probably the best for their needs. Another is that it gives
the researchers who develop such systems a clear goal to work toward and a way to
determine their progress relative to that of other researchers. It is in this final role
that good benchmarks can help to stimulate a research community to produce more
and better results where the quality of the results is well-understood.

They can, in fact, turn research into a competition with clearly defined rules
where it is possible to objectively beat earlier research. However, this obviously
only works well if the standardized tests indeed measure the right properties and
measure them well. It is therefore a crucial property of a benchmark that it is widely
accepted by the relevant research community as having these properties.

There are different ideas about what makes a benchmark the most effective as a
tool to drive research forward. On the one hand, it might be argued that a benchmark
should be *focused* in the sense that it concentrates on particular bottlenecks, for
example, particular features in certain queries that are hard to implement efficiently,
for which the community would like to make progress. Such benchmarks typically
consist of a small number of fixed data sets and workloads, or even just parts of
workloads such as in *microbenchmarks*. On the other hand, it can be argued that a
benchmark should be *broad* in the sense that it attempts to cover a whole class of
workloads that might appear in practice. Such benchmarks typically have generated
data sets and workloads. Next to this dimension, a benchmark can also be narrow
or broad in the sense that it is more or less *domain-specific*. As is also argued in
the introduction of Gray (1993), even though many successful benchmarks in the
past, such as the TPC benchmarks, have been focused, there are both advantages
and disadvantages to benchmarks that are focused or broad in these two senses.
For example, a focused benchmark gives a clear focus to research with well-
understood goals for what are considered the hard problems and promotes small
feasible advances for them. In contrast, broad benchmarks encourage out-of-the-
box thinking, can more often be used in unforeseen applications, and discourage
tweaking of solutions that toward certain artifacts of the benchmark. We will therefore take in this work the position that all these types of benchmarks are relevant.

In this chapter, we restrict ourselves to benchmarks for graph data management,
specifically for graph databases and graph-processing frameworks. We exclude
here benchmarks for related systems such as NoSQL databases and distributed
data-processing frameworks that do not specifically and explicitly address graph
processing. However, the benchmarks we do consider still vary greatly in several
aspects:

System Under Test: The System Under Test (SUT) might be stand-alone or it
    might be part of a distributed system where also other processing happens. The

SUT might be considered as a black box, or it might be considered as consisting of components of which we also would like to measure the performance.

Benchmark process: The benchmark process describes how the system and its context are set up for the measurements and how the measurements are done. This might consist only of loading the data and executing the predefined workload, but it might also include starting up other workloads that run in parallel. For database benchmarks, it might also include rules about repeating the measurements under different circumstances to take for example caching into account.

Data sets: The data sets that need to be stored, retrieved and processed might be fixed or generated. If they are generated it might be that all kinds of aspects and characteristics of the data that influence the measurement might be varied in a controlled way to make sure that the data are representative. For graph data this typically includes graph properties such as the distribution of the degrees of the nodes, but also the distribution of average distance between nodes, or for example the distribution of clusters of highly connected nodes.

Workload: The workload for graph-processing frameworks might consist of relatively simple computational tasks or algorithms, or complex workflows consisting of such tasks, or even of collections of such workflows that all should be executed in parallel. For databases the workload might for transaction processing consist of collections of transactions, containing usually both updates and queries, and specification of the frequency of these transactions. If the benchmark focuses on query processing, the workload might simply be collection of queries, which might be fixed or generated according to some specifications. For graph processing the workloads will typically contain mostly operations that concern the graph structure in the data such as computing certain graph properties and graph queries such as path expression.

Measurements: The measurements can in the simplest case just be the time it takes on average to complete the workload, but it can also be the consumed computational resources such as memory, number of CPUs, CPU time and network bandwidth. Alternatively, the measurement might be aimed at determining the quality of the output, for example, in the cases of community detection algorithms, entity matching algorithms and schema mapping algorithms. In those cases the measurement usually consists of a comparison to a predefined ground truth that defines the ideal output.

All the mentioned aspects of the benchmark are usually influenced by to what extent the benchmark is *domain oriented* and what this domain is. In different domains we find networks with different characteristics and different computational tasks that need to be performed on these networks. Therefore, the data sets and workloads that might be considered typical might be different. Examples of such domains might be biology, the life sciences, sociology, finance, the web, social networks, telecommunication, software engineering, e-commerce, crime fighting, and fraud detection. On the other hand, some benchmarks focus entirely on particular computational tasks on graphs such as finding shortest paths, graph clustering, computing betweenness centrality, finding triangles, and computing Page Rank.

6.2 Survey of Main Benchmarking Approaches

In this section, we survey the main benchmarking approaches. Across the features described in the previous section, we survey benchmarks for RDF databases, benchmarks for graph databases, benchmarks for parallel and distributed graph-processing systems, and data-only benchmarks, in turn. We take a balanced view across the domain, and ascribe the differences in the number of surveyed benchmarks for each benchmark type to the historical maturity of the fields when the first benchmarking approaches were proposed, and to the diversity of system architectures appearing in each field.

6.2.1 RDF Databases

The Resource Description Framework (RDF) and RDF Schema (RDFS) are W3C standards for modeling and sharing graph-structured linked data on the web (Cyganiak et al. 2014; Brickley and Guha 2014). RDF represents relationships between web resources using triples of the form (*subject, predicate, object*), indicating that there is a relationship from resource *subject* to resource *object* described by resource *predicate*. An RDF graph is then a finite collection of triples. RDFS is a vocabulary for describing, by means of triples, ontological structures in RDF graphs, such as class–subclass hierarchies and domain/range restrictions on predicates. SPARQL is the W3C standard for expressing queries over RDF graphs (The W3C SPARQL Working Group 2013).

RDF, RDFS, and SPARQL have experienced broad uptake in the (Semantic) Web community. As the adoption of these standards has grown so has the development of commercial and open source systems for management of massive RDF graphs. In conjunction with the study and engineering of these so-called triple stores, benchmarks specifically targeting RDF data management have been proposed in the community.

In this section, we survey RDF benchmarks, with a focus on six exemplary proposals (summarized in Tables 6.1 and 6.2). Our goal is to illustrate the challenges and distinguishing features of RDF benchmarks. As our focus here is on graph benchmarking, we will not address the role of special RDF features, such as ontological reasoning and blank nodes, in our survey.

6.2.1.1 Lehigh University Benchmark (LUBM) (Guo et al. 2005)

LUBM was one of the earliest proposals for an RDF benchmark, motivated by rapidly expanding development of Semantic Web technologies. The LUBM framework consists of an ontology modeling a university domain, a data generator

Table 6.1 Characterizing features of existing RDF benchmarks (Part 1)

Name	Input	Output	Type of workload
LUBM	# of universities, random seed	Graph instances	Analytical
BSBM	# of products	Graph instances	Analytical
SP2Bench	# of years/triples, random seed	Graph instances	Analytical
AO	RDF graph, # of triples degree of structuredness	Graph instances	–
RBench	RDF graph, scale factors	Graph and workload instances	Analytical
Grr	Graph schema descriptions	Graph instances	–

Table 6.2 Characterizing features of existing RDF benchmarks (Part 2)

Name	Data model	Query language support	Distinguishing features, choke points
LUBM	RDF	SPARQL	Large input, low output queries
BSBM	RDF	SPARQL, SQL	Query mixes
SP2Bench	RDF	SPARQL, SQL	Full coverage of SPARQL features
AO	RDF	–	Data-driven application-specific graph instances
RBench	RDF	SPARQL	Data-driven application-specific graph instances
Grr	RDF	–	Schema-driven application-specific graph instances

for creating synthetic instances of the ontology, and a set of fixed queries for benchmarking query-processing solutions.[1]

The LUBM ontology Univ-Bench models universities, their departments, the people that work in departments, and university activities. The ontology consists of 43 classes (e.g., Professor is a subclass of Faculty, Faculty is a subclass of Employee, and Employee is a subclass of Person; Student is a subclass of Person) and 32 properties (e.g., a Professor can be an Advisor of a Person).

Data generation in LUBM is performed by the Univ-Bench Artificial data generator (UBA). Synthetic ontology instances are generated in units of universities (i.e., the smallest instance consists of one university). Class instances are generated randomly using a user-provided seed, with hard-coded lower and upper bounds on

---

[1]http://swat.cse.lehigh.edu/projects/lubm/.

the sizes of classes (e.g., each university has a minimum number and maximum number of departments).

There are 14 fixed queries in the LUBM benchmark, ranging from small queries with no joins to large queries with up to five joins. The design goal of these queries is to provide broad coverage of input sizes (i.e., the proportion of classes and class instances involved in query evaluation), selectivity (i.e., the ratio of query output size to input size), and complexity (i.e., the number of joins and selection conditions). In order to stress-test RDF engines, the LUBM designers particularly biased the benchmark queries toward large input size and high selectivity (i.e., queries with large inputs and relatively small outputs).

In summary, LUBM was the first RDF benchmark to provide a complete principled solution to triple store benchmarking, with a particular stress on data-intensive queries (rather than on ontological reasoning). Consequently, LUBM has had a major impact on the research community, with over 1200 citations at the beginning of 2018.

6.2.1.2 Berlin SPARQL (BSBM) and SPARQL Performance (SP2Bench) Benchmarks (Bizer and Schultz 2009; Schmidt et al. 2009)

Following up on LUBM are the BSBM[2] and SP2Bench[3] benchmarks. Like LUBM, both benchmarks adopt fixed graph schemas and fixed sets of benchmark queries and provide tools for synthetic instance generation. Furthermore, while the focus is on RDF graphs and SPARQL queries, both also provide relational representations of instances and the benchmark queries, (i.e., relation tables and SQL queries).

BSBM models an e-commerce scenario, with eight classes (e.g., Product, Producer, Vendor, Review, and Person) and seven properties between these classes (e.g., Products are Produced by Producers; Products Have Reviews). BSBM data generation centers around Product instances. Given a user-provided scale factor n, BSBM generates n products and assigns values to the properties of each product following fixed production rules (e.g., each product has between 3 and 5 numeric properties, each having a value between 1 and 2000, following a normal distribution). Furthermore, for each product instance, related instances of the other classes such as producer and review are similarly created following fixed production rules.

The instances generated by SP2Bench follow a bibliographic scenario, modeled after the well-known DBLP[4] data set. Here node types include eight types of documents (e.g., Article, Book) and Person, and 22 relationships such as References (from documents to documents) and Creator (from documents to persons). Given a number of years or a number triples to generate, instance generation is based on fixed production rules and a user-provided random seed, just as with LUBM.

---

[2] http://wifo5-03.informatik.uni-mannheim.de/bizer/berlinsparqlbenchmark/.

[3] http://dbis.informatik.uni-freiburg.de/index.php?project=SP2B/.

[4] http://dblp.uni-trier.de/xml/.

The fixed benchmark queries of both BSBM and SP2Bench aim to extend the stress-testing of the LUBM queries with further features of SPARQL and variety of queries. BSBM includes 12 benchmark queries that include place holders for randomly chosen constant values, with the aim of simulating query mixes (i.e., sequences of instantiations of the benchmark queries) typical of consumers searching for particular products. SP2Bench's 17 benchmark queries, on the other hand, have a particular focus on deeper stress testing, data-intensive query processing, beyond LUBM, including for example non-monotonic queries (i.e., those with negation).

To summarize, BSBM and SP2Bench extend the fixed-schema and workload benchmarking approach introduced with LUBM, with new graph and query scenarios and further data-intensive features of SPARQL. Both benchmarks have significantly influenced and shaped subsequent developments in the field, with BSBM and SP2Bench each having hundreds of citations already in early 2018.

6.2.1.3 Data-Driven RDF Benchmarks: Apples & Oranges (AO) and RBench (Duan et al. 2011; Qiao and Özsoyoglu 2015)

We next discuss a pair of benchmarks that take a completely different approach from the fixed-schema fixed-workload LUBM, BSBM, and SP2Bench benchmarks. Motivated by the observation that these three RDF benchmarks generate instances that differ markedly from real-world RDF graphs (Duan et al. 2011), the Apples & Oranges (AO) benchmark proposes a data-driven approach to synthetic instance generation. In particular, given an RDF graph instance (i.e., a "seed"), a desired output instance size, and a desired level of "structuredness" (i.e, how regular the graph should be), AO generates a synthetic graph instance of the target size and structuredness, mimicking the characteristics of the given seed input graph. In this way, users can finely control and tailor benchmarks instances to a given application, unlike with the fixed-schema benchmarks.

Continuing in this data-driven approach to benchmark instance generation, RBench provides for further control of the generated synthetic instance (in terms of scale factor and degree factor) and for synthetic query workload generation. Generated queries can be shaped as chains, stars, trees, and cycles. The structures and edge labels occurring as basic building blocks of queries are selected from the frequent structures extracted from the input graph instance, which are also used in graph generation. Richer structures, such as trees and cycles, are then formed by random walks on the graph instance.

Together, AO and RBench provide users with the ability to generate sophisticated tailored benchmarks that exhibit the rich structure of real RDF graphs.

6.2.1.4   Generating Random RDF (Grr) (Blum and Cohen 2011)

In a sense, AO and RBench are not strictly benchmarks, but rather tools for generating benchmarks tailored to particular application domains. We close our discussion of RDF benchmarks by briefly noting an additional interesting approach to synthetic instance generation, Grr, which could be used as a basis for benchmarking triple stores, which does not take a data-driven approach.

Grr is a system for generating random synthetic data, which explicitly presents itself *not* as a benchmark, but rather as a tool for data-driven application testing (i.e., a tool for generating concrete benchmarks). This system provides an abstract declarative language and programming model for specifying essentially graph schemas for synthetic data generation. In particular, using a SPARQL-based syntax, users can define construction patterns that define nodes and edges to generate, built using user-provided procedures to generate data values appropriate for the domain being modeled. The framework is demonstrated to be effective in generating large graphs matching the LUBM scenario in addition to the well-known FOAF schema. Furthermore, the intuitiveness and ease-of use of the Grr language is richly demonstrated with these scenarios.

Grr's flexible, controlled, schema-driven approach to graph instance generation distinguishes it from the other five RDF benchmarks surveyed here.

6.2.2   Graph Databases

Graph databases emphasize queries that are quite different from relational database queries, for which well-established benchmarks (Gray 1993; Transaction Processing Performance Council (TPC) 2016) exist. In particular, graph queries may involve both bounded and unbounded recursion, may also exhibit complex patterns and aggregates, or may entail expensive analytical operations, such as PageRank, centrality, or clustering. Each of these query types motivates specific benchmarking needs, none of which can be found in other benchmarks for graph-like data models, as for example in benchmarks for XML (Schmidt et al. 2002; Yao et al. 2004; Barbosa et al. 2009) or for Object Oriented Databases (Cattell and Skeen 1992; Carey et al. 1993). In this section, we summarize leading benchmarking efforts in graph data management (summarized in Tables 6.3 and 6.4).

The Linked Data Benchmark Council (LDBC)[5] (Erling et al. 2015) is an industry-neutral foundation for developing graph and RDF benchmarks, auditing and publishing benchmark results. The benchmarks conceived by LDBC allow to quantitatively compare different technological solutions, helping IT users to make more objective choices for their software architectures and to stimulate technological progress among graph data management systems. The first LDBC

---

[5]http://www.ldbcouncil.org/.

Table 6.3 Characterizing features of existing graph database benchmarks (Part 1)

Name	Input	Output	Type of workload
LDBC	Scale factor	Graph instances/queries updates	Analytical/transactional
WatDiv	Scale factor/fixed schema, fixed query templates	Graph instances/queries	Transactional
gMark	Scale factor, arbitrary schemas, bidirectional distributions	Graph instances/queries	Transactional
GSCALER	Graph instance, Scale factors	Graph instances	–

Table 6.4 Characterizing features of existing graph database benchmarks (Part 2)

Name	Data model	Query language support	Distinguishing features, choke points
LDBC	Arbitrary	Arbitrary	Choke-point-driven design, cardinality estimation, join order, parameter curation, parallelism
WatDiv	RDF	SPARQL	Schema-driven generation, query workload diversity
gMark	Directed edge-labeled graph	Arbitrary	Schema-driven selectivity estimation, recursive queries, diversity control, user-defined schemas
GSCALER	Directed graph	–	Data-driven application-specific graph instances

benchmark is the social network benchmark (SNB) that simulates user activity in a social network. LDBC benchmark design is guided by the notion of a choke point, which is an aspect of query execution or optimization that is known to be problematical for the present generation of various DBMS (relational, graph and RDF). SNB includes a data set generator (DATAGEN) and a set of complex queries. The data set generation relies on a fixed schema consisting of 11 entities (such as Persons, Tags, Forums, Messages, Likes, Organizations, Places, etc.) and 20 relations (such as knows, studyAt, workAt, likes, hasMember, etc.). The schema of SNB encodes the associations between the different entities via standard UML notation (such as one-to-many or many-to-many associations) and leverages subtyping (e.g., City, Country, and Continent are all subtypes of the type Place). Another important feature of the instance generation performed in DATAGEN is the ability to produce a highly correlated social network graph, in which attribute values are correlated among themselves and also influence the connection patterns in the social graph. As an example, the place where a person was born and gender influence the first name distribution. The volume of person activity in a real social network, that is, number of messages created per unit of time, is not uniform, but rather driven by spiking trends. Therefore, an event-driven post generation can be enabled to generate spikes of different magnitude. Structure correlation is also taken care of since the number of friendship edges generated per person (friendship degree) is skewed. Parameter Curation (Gubichev and Boncz 2014) is an important choke point of DATAGEN that allows to find substitution parameters with equivalent behavior via a data mining step during data generation.

DATAGEN can generate social networks of arbitrary size, by varying the input scale factor, which corresponds to the amount of GB of uncompressed data in comma separated value (CSV) representation. DATAGEN can also generate RDF data in Ntriple format. DATAGEN is implemented on top of Hadoop to provide scalability.

The SNB-interactive workload consists of three query classes: *transactional update queries* that can be generated by using DATAGEN by following some fixed query patterns; *simple read-only queries*, consisting of simple lookups; *14 complex read-only queries*, involving friendship patterns, and other related social queries presenting the core of query optimization choke points in the benchmark. The LDBC web site provides query definitions in SPARQL, Cypher, and SQL, as well as API reference implementations for Neo4j and Sparksee.

Among the choke points stressed by the SNB-interactive query workload, we mention: (1) *the estimation of cardinality in transitive queries*; (2) *identifying the right join order and type*, which is more challenging for triple patterns than for relational queries; (3) *handling scattered index access patterns* to allow graph traversals (such as neighborhood lookup) with random access and without predictable locality; (4) *parallelism and result reuse*, by ensuring query workloads with intra- and inter-query parallelism. Other two query workloads of SNB, namely, the Business Intelligence and Graph Algorithms, are currently under development.

Summarizing, among the novel features of SNB, the choke-point-driven design, the user and expert inputs in the design of the domain, and the presence of mixed query workloads, including updates are worth mentioning.

The Waterloo SPARQL Diversity Test Suite (WatDiv)[6] (Aluç et al. 2014) supports user-defined schemas via a so-called data set description language and employs them in the graph generation. Precisely, WatDiv supports local constraints for the graph predicates. As an example, the local constraints in WatDiv specify for instance that 30% of the products have a content rating. For what concerns the graph encoding, WatDiv relies solely on RDF, even though it also makes use of subtyping similar to LDBC (e.g., a product can be an album, a movie, a concert, and so on). Regarding the query workload, WatDiv introduces two classes of query features,

---

[6]http://dsg.uwaterloo.ca/watdiv/.

namely, structural features and data-driven features that should be used to evaluate the diversity of the datasets and workloads in a SPARQL benchmark. The key notion is a constrained basic graph pattern (CBGP), which is a finite set of triple patterns and a finite set of SPARQL filters. Then, structural features defined upon CBGPs are for instance the triple pattern count or the join vertex count, the first representing the number of triples involved in a query and the latter representing the number of join vertices across multiple triple patterns. Then, data-driven features of the queries in the workload are the result of cardinality and the selectivity of CBGPs.

WatDiv consists of multiple tools capable of stress testing RDF data management systems. The data generator generates scalable data sets at user-specified scale factors. Data are generated according to a fixed WatDiv schema with customizable value distributions. The query template generator traverses the WatDiv schema and generates a diverse set of query templates. Users can specify the number of query templates to be generated as well as other parameters on the query templates such as the maximum number of triple patterns or whether predicates in triple patterns should contain constants. Given a set of query templates, the query generator instantiates these templates with actual RDF terms from the data set. The number of actual queries to be instantiated per query template can be specified by users. Given a WatDiv data set and test workload, for each query in the workload, the feature extractor computes the structural and data-driven features discussed above. To this purpose, the tool needs to point to a third-party RDF data management system. Summarizing, WatDiv features a data generator based on a schema and a query generator based on a template. This allows for some degree of freedom by customizing distributions in the graph, making Watdiv more diverse than other benchmark tools; however, query generation requiring a data set instance is not entirely schema-based.

gMark, a graph database benchmark (Bagan et al. 2017; van Leeuwen et al. 2017) is the first domain- and query-language-independent synthetic graph benchmarking tool. The advantage of using gMark compared to other benchmarks is that it leverages a flexible graph schema that allows to model any domain, and that it allows to generate queries based on this schema instead of requiring an instance of the graph. Other important features of gMark are the coverage of regular path queries (RPQs) and the possibility of generating query workload with given schema-driven selectivity estimation, as explained next.

The input for gMark is an XML file called the schema definition. It consists of two parts, out of which the first part sets constraints during graph generation by the following variables: types and predicates (expressed as proportions or fixed numbers); for each predicate, in and out degree distributions; while the second part describes the (possibly multiple) query workloads on a schema, by specifying the properties the queries should have and the number of queries to be generated. Within the second part, one can set the following parameters of a query workload: its size (number of possibly distinct queries to be generated); minimum and maximum number of conjuncts (or disjuncts) of queries; maximum and minimum path length; recursion, namely, the probability (between 0 and 1) for a query to contain a Kleene-star operator; the arity of the query, namely, the number of variables in the query head; and the selectivity. The latter parameter determines the expected number of results by a query on a graph, by choosing among three classes: constant (the number of output results stays steady as the graph increases in size); linear (the number of results grows at roughly the same rate as the graph does); quadratic (growth of results is quadratic to graph growth).

gMark is designed to be language-independent and broadly applicable, so the instance data can be expressed as CVS, customized format on a given graph database, or N-Triples, and so on. Similarly for the generated queries, the supported fragment is the Union of Conjunctive Regular Path Queries, an important fragment of graph queries (Pérez et al. 2010). The desired syntax can be obtained depending on the system one wants to test (e.g., SPARQL, LogicQL, OpenCypher, and recursive SQL are those supported in the current release[7]).

Summarizing, gMark pursues even further the diversity of query workloads initiated by WatDiv, by relying on a more expressive language fragment and also by letting specify the desired selectivity of the queries, the latter feature being not present in previous benchmarks. Another notable feature is the schema-driven generation, which guides both instance and query workload generation, making the latter independent of the underlying graph instance.

GSCALER is a graph database generator (Zhang and Tay 2016). This recent approach to synthetic graph generation is in the same spirit as AO and RBench. Following a novel approach inspired by shotgun sequencing in DNA analysis, the generator takes an example graph as input and produces a synthetic graph instance exhibiting similar structure to that of the input graph. The synthetic instance can be both scaled down and scaled up in size. Complementing AO and RBench, GSCALER targets preservation of basic graph properties such as degree and community structure. Experimental study demonstrates the high quality of generated instances, relative to state-of-the-art approaches.

6.2.3   Parallel and Distributed Graph-Processing Systems

We survey in this section benchmarks developed for parallel and distributed graph-processing systems. We focus on key features such as input, output, workload type, and data model. We also focus on the SUT, which can range from typical distributed and HPC systems, to more modern multicore, and GPU-based or CPU+GPU systems. We also identify for each benchmark the distinguishing features, and analyze the history of main ideas and the influence of important benchmarks on future technology. Tables 6.5 and 6.6 summarize this survey.

We also focus on the data model or structure supported in each benchmark, for which we consider edge lists (EL), where a single input file describes one edge per line; edge-vertex list (EVL), where a first input file is in EL structure, and a second

---

[7]https://github.com/graphMark/gmark.

Table 6.5 Characterizing features of existing benchmarks for distributed and parallel graph-processing systems (Part 1)

Benchmark name	Input	Output	Data model/structure	SUT	Distinguishing features, choke-points
HPC-SGAB (2005–2009)	Scale factor	User-level metrics	Static, (EL)	Parallel	Introduces TEPS
Graph500 (2010–) (HPC-SGAB successor)	Scale, edge factors	User-level metrics	Static, (EL)	Parallel	De facto standard HPC community
GreenGraph500 (2014–) (Graph500 extension)	Scale, edge factors	User-level metrics	Static, (EL)	Parallel	Focus on energy consumption
Traversal operations (2012)	Scale factor (number of vertices)	User-level metric	n/a	Single-node	Property graphs
WGB (2013)	Number of seed characters (k) probability of separator (q)	n/a	n/a	Various (distributed)	Programming models
Early benchmark (2013)	None (not scalable)	User-level metric (runtime)	Static (unclear which)	Distributed, multicore systems	Distributed vs. multicore
CloudSuite (2012–2016)	None (not scalable)	Performance counters, runtime	Static (unclear which)	Multicore systems	Multicore, hardware performance counters
proto-Graphalytics (2014–2015)	Scale factor (only for synth. data)	User- and system-level metrics	Various, static (EVL, EVLP)	Various (distributed, CPU+GPU, single-node parallel)	Algorithm coverage, weak and strong scalability
LDBC Graphalytics (2015–)	Scale factor (only for synth. data)	User- and system-level metrics, graph instances in validation mode	Various, static (EVL, EVLP)	Various (distributed, CPU+GPU, single-node parallel)	Industrial-grade, comprehensive process
GraphBIG (2015)	Scale factor (only for synth. data)	User- and system-level metrics, graph instances in validation mode	Dynamic (CSR/COO)	Various (distributed, CPU+GPU, single-node parallel)	Dynamic graphs, modeling IBM System G platform

Table 6.6 Characterizing features of existing benchmarks for distributed and parallel graph-processing systems (Part 2)

Benchmark name	Complexity	Dataset	Load	Programming model
HPC-SGAB (2005–2009)	Kernels (operations, algorithms)	Synth., R-MAT	BFS, classify, betw.cen., construction	Generic
Graph500 (2010–) (HPC-SGAB successor)	Kernels (operations, algorithms)	Synth., R-MAT	BFS, construction	Generic
GreenGraph500 (2014–) (Graph500 extension)	Kernels (operations, algorithms)	Synth., R-MAT	BFS, construction	Generic
Traversal operations (2012)	Query, kernel (algorithm)	Synth., LFR	1 online query, 1 iterative kernel	Generic
WGB (2013)	Kernels (algorithms)	Synth., RTG (Akoglu and Faloutsos 2009)	5 online + 6 update queries, 3 iterative kernels	Generic
Early benchmark (2013)	Kernel	Real, from SNAP	1 core algorithm (k-core decomp.)	Generic
CloudSuite (2012–2016)	Kernel	Real, from Twitter	1 core algorithm (PageRank)	Generic
proto-Graphalytics (2014–2015)	Kernels (algorithms)	Synth. + real	5 core algorithms	Generic
LDBC Graphalytics (2015–)	Kernels (algorithms)	Synth. + real	6 core algorithms, several noncore	Single job
GraphBIG (2015)	Kernels (algorithms)	Synth. + real	5 core algorithms, several noncore	Single job

file includes all possible vertex identifiers (possibly sorted); edge and vertex lists, with properties (EVLP), where an EVL file may additionally include columns for properties, and so on.

HPC Scalable Graph Analysis Benchmark (HPC-SGAB) (Bader and Madduri 2005) is the first benchmark proposal for parallel (HPC) systems running graph-processing workloads. Formulated as a general specification in 2005, it has been refined by a committee of authors and published in full form around 2009. This seminal benchmark, which later has been followed by the very popular Graph500 benchmark, proposes in its 2009 form (Bader et al. 2009) a scalable data generator using the R-MAT power-law graph generator and a scale factor as parameter, four workload kernels of which BFS and a graph-processing algorithm for computing betweenness centrality, and introduces TEPS as the key metric to assess the performance of the system.

Graph500 (Bader et al. 2010) is the successor of HPC-SGAB, whose specification it clarifies and tightens. Graph500 focuses on two kernels, of which one generates the graph and the other is the execution of the BFS algorithm. Only the latter kernel is timed, and the result is reported in TEPS. Graph500 is currently the de facto standard for the HPC community, with the results archive at http://www.graph500.org/ presenting results for over 200 systems (last update, June 2016).

GreenGraph500 (Hofler et al. 2014) is an extension of Graph500 that focuses on energy consumption. GreenGraph500 focuses on the same kernels and has the same elements as Graph500. The main performance metric in TEPS/W. GreenGraph500 is currently the de facto standard for the HPC community, with the results archive at http://green.graph500.org/lists.php presenting results for over 35 systems (last update, June 2016).

A benchmark focusing on traversal operations (Ciglan et al. 2012) repeats the contribution of HPC-SGAB in the databases community, but its key contribution is the focus on property graphs. The benchmark focuses on BFS, which it restricts to 2- and 3-hop operations, and uses a scalable generator for synthetic data sets (LFR). To the best of our knowledge, this benchmark has not been extended and is no longer maintained.

WGB (Ammar and Özsu 2013) is an ambitious proposal for a universal graph benchmark. A key innovation that will last is the focus on different programming models, here, the Hadoop-based MapReduce and the Pregel-like vertex-centric. The core workload consists of several online, update, and iterative queries; the iterative queries (algorithms) are PageRank and an algorithm for clustering. The proposal includes data generation using a synthetic generator based on the RTG algorithm, which it evaluates experimentally. The benchmark does not propose a specific performance metric and does not specify a data model. To the best of our knowledge, the queries were never implemented or used to test real-world parallel and/or distributed systems, and the benchmark is no longer maintained.

An early study (2013) of multiple distributed graph-processing systems (Elser and Montresor 2013) is the first to compare distributed and multicore parallel systems for graph processing. This benchmark uses a generic specification coupled with platform-specific code to implement a core graph-processing algorithm. The benchmark uses real-world data sets from the SNAP archive (see Sect. 6.2.4); thus, the input is realistic, but the benchmark is not scalable. The benchmark uses runtime as its main performance metric. The authors conduct experiments on five different systems.

CloudSuite ('15 extension) (Ferdman et al. 2012) takes a similar approach as the previous benchmark in its 2015 extension with graph analytics (http://cloudsuite. ch/graphanalytics/). It focuses on PageRank, a graph analytics kernel popularized by Google's search business, applied to one real-world data set published in a previous study about Twitter. The key innovation of this benchmark is the use of low-level, hardware performance counters, which leads to a seminal analysis of architectural impact on the performance of multicore systems running graph-processing workloads.

Proto-Graphalytics (Guo et al. 2014, 2015) is a series of benchmarking prototypes that, much like HPC-SGAB earlier for Graph500, will stimulate the community to form around the creation of de facto standard for benchmarking distributed systems running graph-processing workloads. The first study published on proto-Graphalytics (2014) introduces a process for selecting *multiple* core kernels with provable algorithm coverage, real-world and a scalable synthetic data generator derived from Graph500, and a focus on weak and strong scalability to characterize performance. This study also leverages best practices from the field, focusing on five kernels, user- and system-level metrics. The later extension (2015) introduces for the first time capabilities to benchmark GPU and hybrid CPU+GPU systems, next to multicore systems. This work has generated much follow-up work. For example, the experimental study of Pregel-like systems (Han et al. 2014) proposes as a possible key contribution the use of a star-rating system for a variety of criteria; however, the attribution of stars is not quantified, and thus cannot be reproduced in practice. As another example, an experimental study focusing on programming models (Satish et al. 2014) proposes native implementations for a variety of algorithms and platforms; the authors focus on manually optimizing the implementations and thus are able to assess the limits of diverse programming models. Last, another experimental study using more algorithms (Lu et al. 2014) and graphs gives more insight into the interplay between workload characteristics and performance, but does not formulate a performance model.

LDBC Graphalytics (Capota et al. 2015; Iosup et al. 2016) is the first industrial-grade specification of a graph analysis benchmark. LDBC Graphalytics focuses on a comprehensive benchmarking process that reports TEPS, scalability, and robustness metrics. It introduces support for property graphs, and uses a scalable and configurable data generator that can represent Facebook-like social-network graphs with much more realistic properties than Graph500 (Erling et al. 2015). For the industrial focus, the benchmark includes a renewal process, allows for vendor-implemented drivers, and uses modern software engineering practices (continuous development/continuous integration, low technical debt, support for many data formats, and release of harness and exemplary drivers). LDBC Graphalytics also refines the ideas introduced by proto-Graphalytics, for example, by extending the workload selection process, adding more algorithms, and introducing an extensive validation process for each supported algorithm, including epsilon and equivalence matches, which do not require the output to be identical to a prescribed solution, in both format and values. This benchmark is since 2016 supported by the LDBC consortium (Angles et al. 2014) for benchmarking linked data (www.ldbcouncil. org).

GraphBIG (Nai et al. 2015) Developed for modeling the IBM System G platform, GraphBIG provides a valuable extension of work on hardware performance counters (CloudSuite) and comprehensive process (Graphalytics): the inclusion of dynamic (mutable) graphs in the workload. To the best of our knowledge, this benchmark has not received an update since 2015.

6.2.4 Data Sets Used for Benchmarking

Community resources for graph benchmarking also take the form of real-world data sets and data collections. In the following, we give an overview of some of the more popular data sets and data collections that are freely available and often used for benchmarking.

6.2.4.1 SNAP Data Sets/Stanford Large Network Dataset Collection

Location: https://snap.stanford.edu/data/
Origin: This repository of data Sets started in 2004 and grew from research in analysis of large social and information networks. The data Sets available on the web site were mostly collected (scraped) for the purposes of that research. The web site for the collection was launched in July 2009.
Interesting sets: The largest network that was analyzed so far using the library was the Microsoft Instant Messenger network from 2006 with 240 million nodes and 1.3 billion edges. The collection also contains social network graphs with ground truth describing communities.
Managers: The *Stanford Network Analysis Project* at Stanford University.
Domains: Social networks, Communication networks, Citation networks, Collaboration networks, Web graphs, Co-purchasing networks, P2P networks, Road networks, Online communities, Online reviews, Twitter, and Blogs

6.2.4.2 KONECT: The Koblenz Network Collection

Location: http://konect.uni-koblenz.de/
Origin: KONECT is a project to collect large network data sets to support research in the area of network mining. The site of the collection also provides statistics and plots, and code to generate all network data sets from the web.

Interesting sets: Twitter graphs of followers with 1.47B edges and 1.96B edges, and a Friendster graph of friends with 2.59B edges.

Managers: Institute of Web Science and Technologies at the University of Koblenz–Landau.

Domains: Authorship, Web commerce, Communication, Tag clouds, Biology, P2P networks, Ratings, Bibliographies, Semantic Web, Social networks.

6.2.4.3   The Web Data Commons

Location: http://webdatacommons.org/hyperlinkgraph/index.html

Origin: The Web Data Commons project was started by researchers from *Freie Universität Berlin* and the *Karlsruhe Institute of Technology* (KIT) in 2012. The goal of the project was to facilitate research and support companies in exploiting the information on the web by extracting structured data from web crawls and provide these data for public download.

Interesting sets: The 2012 graph is claimed to be the largest hyperlink graph that is available to the public outside companies such as Google, Yahoo, and Microsoft. It covers 3.5 billion web pages and 128 billion hyperlinks between these pages. In addition, there are general extracted RDF data sets based on RDFa, microdata, and microformat as found in the crawled data.

Managers: The WDC Project is mainly maintained by the *Data and Web Science Research Group* at the *University of Mannheim*.

Domains: Web pages.

6.2.4.4   The Yahoo Webscope Program

Location: https://webscope.sandbox.yahoo.com/

Origin: The Yahoo Webscope Program is a reference library provided by Yahoo and consists of interesting and scientifically useful data sets for noncommercial use by academics and other scientists.

Interesting sets: It has a particular set of 3.5 TB (uncompressed) aimed at large-scale machine learning that consists of anonymized user interaction data. It contains interactions from about 20 million users from February 2015 through May 2015, including those that took place on the Yahoo homepage, Yahoo News, Yahoo Sports, Yahoo Finance, and Yahoo Real Estate. The data set contains user-connected information like age range, gender, and some geographic data. There is also news-item information such as title, summary, and key phrases of the news article in question, plus local timestamps.

Managers: Yahoo Labs, the research department of Yahoo.

Domains: Advertising, Market Data, Competition Data, Computing Systems, Social Data, Image Data.

6.2.4.5   The Game Trace Archive

Location: http://gta.st.ewi.tudelft.nl/
Publications: There is one key publication on the data format and original
contents (Guo and Iosup 2012), and further characterization and exploration of
the data (Iosup et al. 2014; Jia et al. 2015).
Origin: The Game Trace Archive (GTA) (Guo and Iosup 2012) was designed
to provide a virtual meeting space for the game community to exchange online
gaming traces. It also defines a unified Game Trace Format (GTF) to facilitate
the exchange of game traces and tools to convert from and to this format. A
key feature of the data sets included in the archive is that they represent basic
relationships between online players, which need to be combined to express as
interaction graphs (Wilson et al. 2012) the more complex pro-social relationships
that emerge in gaming communities.
Interesting sets: There are data sets of different sizes, from hundreds of thousand
edges to tens of millions of edges. The sets cover card games, board games,
MMOFPSs (first person shooters) and MMORPGs (role playing games).
Managers: The *Parallel and Distributed Systems* research group at TUDelft.
Domains: Online gaming.

6.2.4.6   The Billion Triples Challenge Data Sets

Location: http://km.aifb.kit.edu/projects/btc-2014/ (links also to the older data
sets)
Publications: In Käfer and Harth (2014) an overview of the 2014 data set is
presented.
Origin: The datasets were created as a basis for submissions to the Big Data
Track (formerly the Billion Triples Track) of the *Semantic Web Challenge*. There
were sets created in the years 2009, 2010, 2012, and 2014 by crawling for RDF
data on the web. The data set consists of 15 RDF graphs that were each separately
crawled during February to June 2014.
Interesting sets: The sizes of the individual graphs range from tens of thousands
of quads to more than a billion quads. The total sum of quads exceeds 4 Billion.
Managers: The *Web Science und Wissensmanagement* group at the Karlsruhe
Institute of Technology.
Domains: The crawled data sets cover a very wide range of domains, such and
in some sense all domains that the crawl happened to arrive at. This included
domains such as the life sciences, chemistry, national government, scientific
bibliographies, geography, geology, national law, world health, national health,
museums, Nobel prizes and world economics. Alongside, it also included general
data sets such as DBpedia and YAGO3.

6.2.4.7   The Microsoft Academic Graph

Location: http://research.microsoft.com/en-us/projects/mag/
Publications: In Sinha et al. (2015) an overview is given.
Origin: The Microsoft Academic Graph is an entity graph concerning scientific publications that serves as the backbone of Microsoft Academic Service (MAS), which is maintained by Microsoft to support the Microsoft Academic search engine that has been publicly available since 2008 as a research prototype. The graph contains information about publication records, citation relationships between those publications, as well as authors, institutions, journals and conference "venues," and fields of study.
Interesting sets: The single but heterogenous graph contained in 2015 information about more than 83 million papers, 20 million authors, 770 thousand institutions and 22 thousand journals. Since February 2016 it also contains extra files for the KDD Cup 2016.
Managers: The *Microsoft Academic Graph* project is run by the *Internet Services Research Center* (ISRC) of Microsoft.
Domains: Scientific publications.

6.3   Ongoing and Future Work

There is much ongoing work, but still also a wide range of topics for further study concerning the benchmarks discussed in this chapter. We will mention some of them in this section.

New types of performance: As discussed, for example, in the Beckman report presented in Abadi et al. (2016), benchmarks should not only focus on scalability in terms of data size, but also look at metrics such as total cost of ownership, end-to-end processing time (from raw data ingestion to producing insights), and usability for nonexpert users. These all require new types of benchmarks that differ from the existing ones.

Mapping out typical usage within different application domains: The discussed benchmarks are all intended to be general benchmarks, but these systems will be used differently within different domains such as the life sciences, intelligence gathering, sociology, urban analytics, and so on, and have different requirements. It is likely that they will require specialized benchmarks, or that general benchmarks will have to make sure to be relevant for them. One important aspect here, for example, is the interplay and mix between graph-oriented queries, which focus on the structure of the graph, and data-oriented queries, which focus on attribute values, or queries that combine both aspects. What is the typical mix for certain applications, and should this be reflected in the benchmark?

Organizing the evolution of benchmarks: As systems will develop, and also their usage will change, it will be important to define and manage a process to keep benchmarks up-to-date and relevant. Rather than being replaced, benchmarks should be adapted, to prevent duplication of work and the loss of experience. Ideally this should be done in rounds with deadlines, where in each round both commercial and academic systems are evaluated and compared. The deadlines would ensure that relative latecomers do not have an unfair advantage in the competition.

Generation of workflows and graphs: The artificial generation of typical workflows is still largely an open issue, both for static workflows as well as dynamic workflows that evolve in time. The latter requires a better understanding of typical arrival patterns for workloads. For static workflows, there is also still much to do in the area of scalable generation of very large graphs that have the typical and desired combinations of properties, such as the small-world property and being scale-free.

Understanding the relationship between similar or shared components in benchmarks: It will be interesting to create an overview of shared components of benchmarks and how they relate to each other in terms of performance. For example, several benchmarks contain reachability queries, and so it will be important to understand if the performance of that component within one benchmark has a relationship with the performance of similar components in other benchmarks. This overview was not yet attempted in this work, and it would require a common platform on which to run the benchmarks.

We close this section with the remark that this list is of course not exhaustive, but summarizes some of the most important and interesting future research directions. Their successful pursuit will provide important support for both accelerating the developments in the field of graph data management and graph processing, as well as make sure that they remain relevant outside of the research communities.

6.4 Concluding Remarks

In this chapter, we have given an overview of existing benchmarks for graph data management and graph processing as are found in the database research community and the parallel and distributed research community. In the introduction the origin and motivation of benchmarks was discussed, and it was explained how this is perceived in the different communities. Therefore in this section also a common terminology was established, which is used for the presentation and comparison of the different benchmarks discussed in this chapter.

This was followed by a survey of the different benchmarks in the research domains of *RDF Databases*, *Graph Databases*, and *Parallel and Distributed Graph-Processing Systems*. The benchmarks from the database research domain were compared on aspects such as the types of input and output they consume and produced, the type of workloads they consider, the data models or data structures they consider, the query languages that are used, and the distinguishing features and choke points. The benchmarks from the data-processing domain were also compared on aspects such as the type of SUT, the complexity of the considered operations, the origin of the datasets, the considered loads, and finally the programming model. Finally, we also gave an overview of the different collections of graph-oriented data sets that are freely or almost freely available, and that are often used for benchmarking.

At the end of this chapter the main topics for ongoing and future research were presented, to give an idea of the evolution of graph benchmarks in the near future. The benchmarks presented in this chapter illustrate that already much has been achieved. It is interesting to observe that the involved research communities have been slowly converging, as is witnessed by joint papers and participation in each other's events. We hope and expect that this will continue in the future, towards addressing the large shared challenges that still lie ahead.

References

Abadi D, Agrawal R, Ailamaki A, Balazinska M, Bernstein PA, Carey MJ, Chaudhuri S, Chaudhuri S, Dean J, Doan A, Franklin MJ, Gehrke J, Haas LM, Halevy AY, Hellerstein JM, Ioannidis YE, Jagadish HV, Kossmann D, Madden S, Mehrotra S, Milo T, Naughton JF, Ramakrishnan R, Markl V, Olston C, Ooi BC, Ré C, Suciu D, Stonebraker M, Walter T, Widom J (2016) The Beckman report on database research. Commun ACM 59(2):92–99. http://doi.acm.org/10. 1145/2845915

Akoglu L, Faloutsos C (2009) RTG: a recursive realistic graph generator using random typing. Data Min Knowl Discov 19(2):194–209. http://dx.doi.org/10.1007/s10618-009-0140-7

Aluç G, Hartig O, Özsu MT, Daudjee K (2014) Diversified stress testing of RDF data management systems. In: ISWC, pp 197–212

Ammar K, Özsu MT (2013) WGB: towards a universal graph benchmark. In: Advancing big data benchmarks - proceedings of the 2013 workshop series on big data benchmarking, WBDB.cn, Xi'an, July 16–17, 2013 and WBDB.us, San José, CA, October 9–10, 2013 Revised Selected Papers, pp 58–72

Angles R, Boncz PA, Larriba-Pey J, Fundulaki I, Neumann T, Erling O, Neubauer P, Martínez-Bazan N, Kotsev V, Toma I (2014) The linked data benchmark council: a graph and RDF industry benchmarking effort. SIGMOD Record 43(1):27–31. http://doi.acm.org/10.1145/ 2627692.2627697

Bader DA, Madduri K (2005) Design and implementation of the HPCS graph analysis benchmark on symmetric multiprocessors. In: High performance computing - HiPC 2005, 12th international conference, proceedings, India, December 18–21, 2005, pp 465–476

Bader DA, Feo J, Gilbert J, Kepner J, Koester D, Loh E, Madduri K, Mann B, Meuse T, Robinson E (2009) HPC scalable graph analysis benchmark. Online technical specification, v.1.0, Feb 24. http://www.graphanalysis.org/benchmark/GraphAnalysisBenchmark-v1.0.pdf

Bader et al DA (2010) Graph500. Online technical specification, v.0.1 (2010) through 1.2 (2011). http://www.graph500.org/specifications

Bagan G, Bonifati A, Ciucanu R, Fletcher GHL, Lemay A, Advokaat N (2017) gmark: schema-driven generation of graphs and queries. IEEE Trans Knowl Data Eng 29(4):856–869

Barbosa D, Manolescu I, Yu JX (2009) XML benchmarks. In: Liu L, Özsu MT (eds) Encyclopedia of database systems. Springer, Berlin, pp 3576–3579

Bizer C, Schultz A (2009) The Berlin SPARQL benchmark. Int J Semant Web Inf Syst 5(2):1–24

Blum D, Cohen S (2011) Grr: generating random RDF. In: ESWC, pp 16–30

Brickley D, Guha RV (2014) Rdf schema 1.1. W3C recommendation. https://www.w3.org/TR/rdf-schema/

Capota M, Hegeman T, Iosup A, Prat-Pérez A, Erling O, Boncz PA (2015) Graphalytics: a big data benchmark for graph-processing platforms. In: Proceedings of the third international workshop on graph data management experiences and systems, GRADES 2015, Melbourne, May 31–June 4, 2015, pp 7:1–7:6

Carey MJ, DeWitt DJ, Naughton JF (1993) The oo7 benchmark. In: Proceedings of the 1993 ACM SIGMOD international conference on management of data, Washington, May 26–28, 1993, pp 12–21

Cattell RGG, Skeen J (1992) Object operations benchmark. ACM Trans Database Syst 17(1):1–31

Ciglan M, Averbuch A, Hluchý L (2012) Benchmarking traversal operations over graph databases. In: Workshops proceedings of the IEEE 28th international conference on data engineering, ICDE 2012, Arlington, April 1–5, 2012, pp 186–189. http://dx.doi.org/10.1109/ICDEW.2012.47

Cyganiak R, Wood D, Lanthaler M (2014) RDF 1.1 concepts and abstract syntax. W3C recommendation. https://www.w3.org/TR/rdf11-concepts/

Duan S, Kementsietsidis A, Srinivas K, Udrea O (2011) Apples and oranges: a comparison of RDF benchmarks and real RDF datasets. In: SIGMOD, pp 145–156

Elser B, Montresor A (2013) An evaluation study of bigdata frameworks for graph processing. In: Big data

Erling O, Averbuch A, Larriba-Pey J, Chafi H, Gubichev A, Prat A, Pham MD, Boncz P (2015) The LDBC social network benchmark: interactive workload. In: SIGMOD, pp 619–630

Ferdman et al M (2012) Clearing the clouds: a study of emerging scaleout workloads on modern hardware. In: ASPLOS

Gray J (ed) (1993) The benchmark handbook for database and transaction systems, 2nd edn. Morgan Kaufmann, San Mateo

Gubichev A, Boncz P (2014) Parameter curation for benchmark queries. In: TPCTC, pp 113–129

Guo Y, Iosup A (2012) The game trace archive. In: 11th annual workshop on network and systems support for games, NetGames 2012, Venice, November 22–23, 2012, pp 1–6. http://dx.doi.org/10.1109/NetGames.2012.6404027

Guo Y, Pan Z, Heflin J (2005) LUBM: a benchmark for OWL knowledge base systems. J Web Sem 3(2–3):158–182

Guo et al Y (2014) How well do graph-processing platforms perform? In: IPDPS

Guo et al Y (2015) An empirical performance evaluation of gpu-enabled graph-processing systems. In: CCGrid

Han M, Daudjee K, Ammar K, Özsu MT, Wang X, Jin T (2014) An experimental comparison of pregel-like graph processing systems. PVLDB 7(12):1047–1058

Hofler T et al (2014) GreenGraph500. Online technical specification, v.1.1 (2014). http://green.graph500.org/greengraph500rules.pdf

Iosup A, van de Bovenkamp R, Shen S, Jia AL, Kuipers FA (2014) Analyzing implicit social networks in multiplayer online games. IEEE Int Comput 18(3):36–44. http://dx.doi.org/10.1109/MIC.2014.19

Iosup A, Hegeman T, Ngai WL, Heldens S, Prat-Pérez A, Manhardt T, Chafi H, Capota M, Sundaram N, Anderson MJ, Tanase IG, Xia Y, Nai L, Boncz PA (2016) LDBC graphalytics: a benchmark for large-scale graph analysis on parallel and distributed platforms. PVLDB 9(13):1317–1328. http://www.vldb.org/pvldb/vol9/p1317-iosup.pdf

Jia AL, Shen S, van de Bovenkamp R, Iosup A, Kuipers FA, Epema DHJ (2015) Socializing by gaming: revealing social relationships in multiplayer online games. TKDD 10(2):11. http://doi.acm.org/10.1145/2736698

Käfer T, Harth A (2014) Billion Triples Challenge data set. Downloaded from http://km.aifb.kit.edu/projects/btc-2014/

Lu Y, Cheng J, Yan D, Wu H (2014) Large-scale distributed graph computing systems: an experimental evaluation. PVLDB 8(3):281–292. http://www.vldb.org/pvldb/vol8/p281-lu.pdf

Nai L, Xia Y, Tanase IG, Kim H, Lin C (2015) Graphbig: understanding graph computing in the context of industrial solutions. In: Proceedings of the international conference for high performance computing, networking, storage and analysis, SC 2015, Austin, November 15–20, 2015, pp 69:1–69:12

Pérez J, Arenas M, Gutierrez C (2010) nSPARQL: a navigational language for RDF. J Web Semant 8(4):255–270

Qiao S, Özsoyoglu ZM (2015) RBench: application-specific RDF benchmarking. In: SIGMOD, pp 1825–1838

Satish N et al (2014) Navigating the maze of graph analytics frameworks using massive datasets. In: SIGMOD

Schmidt A, Waas F, Kersten ML, Carey MJ, Manolescu I, Busse R (2002) XMark: a benchmark for XML data management. In: VLDB, pp 974–985

Schmidt M, Hornung T, Lausen G, Pinkel C (2009) SP2Bench: a SPARQL performance benchmark. In: ICDE, pp 222–233

Sinha A, Shen Z, Song Y, Ma H, Eide D, Hsu BJP, Wang K (2015) An overview of microsoft academic service (MAS) and applications. In: Proceedings of the 24th international conference on World Wide Web, WWW '15 Companion. ACM, New York, pp 243–246. http://doi.acm.org/10.1145/2740908.2742839

The W3C SPARQL Working Group (2013) SPARQL 1.1 overview. W3C recommendation. https://www.w3.org/TR/sparql11-overview/

Transaction Processing Performance Council (TPC) (2016) TPC benchmark. http://www.tpc.org/

van Leeuwen W, Bonifati A, Fletcher GHL, Yakovets N (2017) Stability notions in synthetic graph generation: a preliminary study. In: EDBT, pp 486–489

Wilson C, Sala A, Puttaswamy KPN, Zhao BY (2012) Beyond social graphs: user interactions in online social networks and their implications. TWEB 6(4):17. http://doi.acm.org/10.1145/2382616.2382620

Yao BB, Özsu MT, Khandelwal N (2004) XBench benchmark and performance testing of XML DBMSs. In: ICDE, pp 621–632

Zhang JW, Tay YC (2016) GSCALER: synthetically scaling a given graph. In: EDBT 2016, pp 53–64

Printed in the United States
By Bookmasters